the edumarketer

Be the marketer of YOUR expertise

Ginger Bell

Copyright © 2019 Edumarketing, LLC., The Edumarketer, Edumarketing Agency, Edumarketing Academy

All rights reserved. No part of this book may be used or reproduced in any manner whatsoever without prior written consent of the author, except as provided by the United States of America copyright law.

Published by Edumarketing, LLC.

Printed in the United States of America.

ISBN: 9781092608633

This publication is designed to provide accurate and authoritative information with regard to the subject matter covered. It is sold with the understanding that the publisher is not engaged in rendering legal, accounting, or other professional advice. If legal advice or other expert assistance is required, the services of a competent professional should be sought.

For more information please write:

Edumarketing
16055 SW Walker Road
Suite 223
Beaverton, OR 97006

Visit us online at: **www.edumarketing.com**

"Always pass on what you have learned."

~ Ginger Bell

I believe that each of us has the ability to share our expertise to help someone else learn from the lessons we have experienced and the knowledge we have gained. It is just a matter of taking the time to share.

Dedicated to The Edumarketer in YOU!

Table of Contents

ACKNOWLEDGEMENT **13**

FOREWORD **17**

USEFUL NOTES **20**

Chapter 1 .... **21**
   THE WHAT AND WHY OF EDUMARKETING *21*

Chapter 2 .... **27**
   QUICK START GUIDE TO EDUMARKETING *27*

Chapter 3 .... **33**
   DISCOVER YOUR EXPERTISE *33*

Chapter 4 .... **39**
   WHO ARE YOUR CUSTOMERS? *39*

Chapter 5 .... **45**
   WHAT PROBLEMS DO YOU SOLVE, WHAT QUESTIONS DO YOU ANSWER? *45*

Chapter 6 .... **51**
   WHAT PRODUCTS/SERVICES DO YOU OFFER? *52*

Chapter 7 .... **63**
   CREATE A CONTENT STRATEGY *63*

Chapter 8 .... **71**
   UNDERSTANDING THE PROCESS *72*

Chapter 9 .... **77**
   HEADLINES COUNT! CREATING AWESOME HEADLINES *77*

Chapter 10 .... **83**
   DEVELOP YOUR CONTENT *83*

Chapter 11 .... **89**
   CREATING MODULAR AND RUBBER BAND CONTENT *89*

Chapter 12 .... **93**
   SELECTING YOUR DELIVERY CHANNELS *93*

Chapter 13 .... **99**
   BUILDING YOUR EDUMARKETING PLAN *99*

Chapter 14 .... **107**
   MARKET, SHARE, PROMOTE AND BUILD *107*

Chapter 15 .... **115**
     FOLLOW-UP AND MEASURING *115*

Chapter 16 .... **121**
   ADJUST YOUR PLAN, CREATE NEW CONTENT AND REPURPOSE *121*

Chapter 17 .... **127**
   SETTING UP YOUR DELIVERY CHANNELS *127*

Chapter 18 .... **147**
   CREATING A CONTENT HUB *147*

Chapter 19 .... **153**
   MANAGING YOUR PLAN *154*

Chapter 20 .... **157**
   HOW TO GET MORE TRACTION WITH YOUR CONTENT *157*

Chapter 21 .... **165**
   HOW TO SET UP AN EDUMARKETING STRATEGY ON A BUDGET *165*

Chapter 22 ....**169**

   KNOWING WHEN AND HOW TO OUTSOURCE..*169*

Chapter 23 ....**175**

   IT'S A MARATHON, NOT A SPRINT – CREATING A LONG-TERM STRATEGY
   ....*175*

Chapter 24 ....**179**

   USEFUL TOOLS, TECHNOLOGY AND RESOURCES...................................*180*

ACKNOWLEDGEMENTS

Special thank you to Theresa Ballard, my dear friend and long-time mentor, who always reminds me to take chances and be fearless.

Thank you to my son, Blaine Bell, who listens to my crazy ideas and helps me reign in on opportunities and to my husband, Blair Bell, who lets me be me and loves me unconditionally.

My ability to create the Edumarketing Academy would not have been possible without the incredible opportunities shared with me by one of my favorite visionaries, Troy McClain. His enthusiasm helped me realize that my gift of creating content could be shared with others. Thank you, Troy, for our years of conversations, coaching and laughter. You've been able to capture my dream of an incredible online social learning platform and I treasure every day I get to work in it.

A heartfelt thank you to my long-time friend and client, Jim Dunkerley, who believed enough in me to let me use edumarketing in his business and begin my journey into creating edumarketing programs that generate results for businesses who embrace the concept and incorporate it into all they do. It truly is a practice that can make a difference. Thank you, Jim, for always believing in the "Unicorn"!

This book is dedicated to my dear friend, Jim Criswell who taught me how to laugh in the face of adversity, love without limitations and live life to it's fullest each and every day. You touched the lives of so many and we will forever be better because of you.

Love you friend!

FOREWORD

On average, Google processes over 40,000 search queries every second. This translates to about 3.5 billion searches each and every day. We have apps like Siri, Google Home, Alexa and Echo that allow us to simply "ask" for information with a voice command. Information is easily available to us, and people are asking for it. Not only are people asking Google for information, but they are also searching for "how to" do things. *Fortune Magazine* reports the top ten "how to" queries on Google are:

1. How to tie a tie
2. How to kiss
3. How to get pregnant
4. How to lose weight
5. How to draw
6. How to make money
7. How to make pancakes
8. How to write a cover letter
9. How to make French toast
10. How to lose belly fat

Besides using Google, people who want to learn by watching rather than reading turn to YouTube, another Google franchise, to search for how to do things. Ask any millennial how they learn how to do things and the resounding answer will be "YouTube."

People are hungry to learn how to do things and this, my friend, is a great opportunity for you! They are searching and **WANT TO KNOW** answers to their questions. So, the question is, what are you doing to help educate them? Do you use your expertise as the foundation of your marketing?

What are you doing to edumarket?

I first used the term "edumarketing" in a workshop for a customer. At the time I didn't know if it was even a word, but what I did know was what I was doing for my customer, and that was taking their knowledge and expertise and using it as a marketing tool in their business. Simple: edumarketing.

I am an Education Specialist. I am not even sure if that is really a job title, although I am sure somewhere it is. For me, I started calling myself an Education Specialist when people kept asking me what I do for my customers. I've been a corporate "trainer" for over 20 years. I've taught sales processes, best practices, how to be a better public speaker, etc. I've "trained" on laws and regulations in the mortgage industry and translated many laws and program guidelines into workshops, presentations, videos, online courses, etc. What I have learned through the years is how to take the information I know, or what my customers know, and turn that information into education and use it to market myself or my customers as experts in their fields. It is edumarketing.

Education + Marketing = Edumarketing

I know I am not the first person to coin this phrase, but I can tell you, I certainly am the person who has used it completely and fully in my business and the businesses of my customers.

I decided it was time to share this business-building gem and teach other professionals how to use edumarketing in their businesses.

My goal in writing this book is to share:

1. The what and why of edumarketing
2. The power of being the expert in your field
3. How to find out what your customers want to know
4. What to do first
5. Various delivery methods and technologies that work
6. How to create your edumarketing plan
7. How to create the right message
8. Where and how to structure your content
9. How to leverage your time
10. Ways to create customer conversions that work
11. How to follow up, stick with your plan, and measure it for effectiveness
12. When you should outsource

Most importantly, my goal is to teach you how to build customers for life using edumarketing!

This book is meant to be the foundation of understanding and creating an edumarketing plan for you and your business. All good foundations begin with a structure. That is what this book is meant to be—a method for you to develop a foundation for your edumarketing plan.

So, let's get started, shall we?

USEFUL NOTES

Chapter 1

THE WHAT AND WHY OF EDUMARKETING

"Make your marketing so useful that people will pay for it."
~ Jay Baer

Before I dive into how to create your edumarketing plan, I think it is important to explain edumarketing.

Edumarketing combines two words:

Education + Marketing = Edumarketing

When we use the word education, we often think of what? School, papers, tests, grades. Yuck!!! But really, education is the transfer of information. The definition of education is the process of receiving or giving systematic instruction. According to our friend Wikipedia[i]:

Education is the process of facilitating learning,

or the acquisition of knowledge, skills, value, beliefs and habits.

Educational methods include storytelling, discussion, teaching, training and directed research. Education frequently takes place under the guidance of educators, but learners may also educate themselves.

Now, let's look at what is marketing. Much more fun than education, for sure!

Marketing is the action of businesses promoting and

selling products or services.

The Business Directory's[ii] definition of Marketing is:

The management process through which goods and services move from concept to the customer. It includes the coordination of four elements called the 4 P's of marketing:

1. Identification, selection and development of a product
2. Determination of its price
3. Selection of a distribution channel to reach the customer's place
4. Development and implementation of a promotional strategy

The fourth element, development and implementation of a promotional strategy, is what we will be focusing on in your edumarketing plan. This is the strategy we will use to educate your prospects, customers and business partners on you, your expertise and how you and your product or service can help them.

For me, I started using edumarketing in my business when I was working at Dale Carnegie Training as a Training Consultant. I had been hired to create a market in the growing field of technology in Portland, OR. I didn't know one single technology person, except my neighbor who worked at Intel. But, what I did know was how to market and build a territory.

The first step in my plan was to create a catchy title for a workshop I would present on mastering public speaking skills. Technology people surely would like to learn how to improve their public speaking, right? I mean, who wouldn't? My catchy title for my workshop was **"Overcoming the Fear of Public Speaking."** After all, out of all the "phobias" out there, it ranks as one of the highest! Glossophobia, which is the fear of public speaking, affects 3 out of 4 people, which translates to nearly 75% of the population. Wow! Seriously?

Ok, so now that I've got your palms sweating because you are probably thinking, there is no way I can use edumarketing in my business because I'm one of those 75%, and there is no way I am going to attempt to conquer that fear, sister! Ok, hold on there, cowboy. I'm going to give you tons, and I mean *tons*, of ideas and ways to use edumarketing in your business without tapping into that phobia of yours...ok, maybe a little; but I promise it will be worth it. I've helped many of my customers overcome the fear of public speaking, or at least provided them with tools they could use so they were able to project the confident professional that they are. We will get into that more later, but now, back to my story.

I had come up with a catchy phrase for my workshop:

"Overcoming the FEAR of Public Speaking"

I wrote up a quick copy for the workshop, and I went to the Executive Director for the local Chamber. I shared with her that the fear of public speaking is one of the biggest fears for most professionals and probably high on the list for many of her chamber members. I shared that I was a Training Consultant with Dale Carnegie Training and that I would be happy to hold a "complimentary" workshop for her members on "Overcoming the FEAR of Public Speaking" if she would like to market it out to her database.

She jumped at the idea, and we set out to hold the first of a series of workshops for the chamber.

My first attempt at using edumarketing in my business was EXTREMELY successful! I had the President of the Oregon Graduate Institute in my workshop along with many other executives from various technology companies in the area. From that one workshop, I was able to sell hundreds of Dale Carnegie courses, including many in-house courses with large companies.

What edumarketing does is allow you to take your expertise;

your specialty;

your ninja power;

and share it with people who want to know how to do what it is that you do or how you can solve a problem that they have. Whatever product you sell, or whatever service you provide, you can edumarket it!

We have a lot to cover, and my goal in writing this book is to give you an easy-to-follow guide that you can use to get your edumarketing plan going right away. After all, what good is learning something if you cannot start using it immediately?

Below are the steps I have used in my own business and the businesses of countless customers to help them take their expertise;

    use it to position themselves as experts in their fields;

        and edumarket the heck out of their business to build some

    impressive results!

Here are the steps:

Step One – Discover Your Expertise

Step Two – Identify What Your Customers Want to Know

Step Three – Create Your Edumarketing Plan

Step Four – Determine Your Delivery Channels

Step Five – Develop Your Content

Step Six – Create Your Message

Step Seven – Deliver Your Message

Step Eight – Follow Up

Step Nine – Measure Your Results

Step Ten – Adjust Your Plan/Create New Stuff

Chapter 2

QUICK START GUIDE TO EDUMARKETING

"These days, people want to learn before they buy, be educated instead of pitched."

~ Brian Clark

You may have picked up this book because you've been told you need to record a video or do something on Facebook or create a podcast about your expertise and you want to get started quickly. I get it, and I've been there. Getting started quickly is important, but it is not how to use edumarketing in your business long-term. Edumarketing is about building a long-term strategy that creates educational marketing programs that will provide education for your prospects and customers and create customers for life. Planning is everything, but you don't have to create your entire edumarketing plan to get started. The information in this chapter is designed to get your edumarketing plan off to a quick start.

Step One: Identify Your Expertise

The first step is to identify your expertise. If you are a financial planner, your expertise is helping people invest, save money, and increase their wealth. If you are a mortgage professional, your expertise is helping people qualify for financing a home of their dreams. Your broad definition of your expertise is great for the local Chamber 30-second elevator pitch, but to successfully build an edumarketing plan, you need to know a bit more. To drill down on your expertise, answer the following questions:

1. What problems do you solve?
2. What questions do you answer?
3. What specialized solutions do you offer?

(Be sure to write down all your answers. These are all possible articles, videos and courses.)

Activity

What Problems Do You Solve?

What Questions Do You Answer?

_____
_____
_____
_____
_____

What Specialized Solutions Do You Offer?

_____
_____
_____
_____
_____

Step Two: Determine Your Topics

Next, take your answers from **Step One** and write down each as a topic. Don't expand on them yet, simply write down each answer. Save this sheet as your topic list. You will find that you can constantly add to this list over time. No matter the question, if you have an answer, you have a topic. When determining your topics, you want to consider who you want to read, view or learn from the topic and what action you want them to take.

Activity

Topics

_____
_____
_____
_____

Step Three: Identify Your Distribution Channels

Distribution is how you deliver your content and engage with your audience. Your distribution channels may be determined by your geographic location or industry but should not be determined by your development expertise. It is easy to outsource to experienced professionals. For example, you may use a ghostwriter to write your content and a videographer to create your videos. If you are looking to develop a complete edumarketing plan that includes online learning, you will want to hire a professional who has a complete understanding for developing educational marketing programs. You don't have to start out utilizing all distribution channels. You can begin by adding blog content to your website and then move on to creating videos and then an eLearning channel. We will cover various distribution channels in more detail later in this book. For now, note a few ideas of what distribution channel(s) you want to begin with.

Activity

Identify Your Distribution Channels

Step Four: Develop Your Content

Start with the following:

1. Take a topic
2. Write out the answer or information to the topic
3. Determine your delivery method(s)
4. Develop your content

You may use the topic and information in a blog post or create a quick informational video to place on YouTube or Facebook. You may create a checklist to use to capture leads on your website or an online course for your eLearning site. You may create a PowerPoint slide deck to use in a webinar or you may use the topic as a discussion in your podcast. You can develop several different deliverables for each topic. The most important thing to remember is to take your topic, write your content, then develop your deliverables. Yes, this process does take a bit of upfront work, but once you have done it, you will be thankful because you will end up with a strong list of topics that will support you in your marketing and business goals and position you as the expert in your field.

Create a content hub to organize your content for your various distribution channels like blog posts, videos, social media, presentations, eLearning, eBooks, etc. If your topic is long, then you may want to create one cohesive paper and/or presentation. Although the content you create may be divided into multiple deliverables, creating one single presentation or whitepaper gives you more control and will help you:

- Divide your broader topics into smaller topics
- Develop multiple content for various distributions channels
- Build FAQ's

Chapter 3

DISCOVER YOUR EXPERTISE

Let's start by understanding the definition of an expert. Wikipedia[iii] gives us the following definitions of an expert:
1. An expert is someone who has a prolonged or intense experience through practice and education in a field.
2. An expert is someone widely recognized as a reliable source of technique or skill whose faculty for judging or deciding rightly, justly, or wisely is accorded authority and status by peers or the public in a specific well-distinguished domain.
3. An expert is a person with extensive knowledge or ability based on research, experience, or occupation and in an area of study.
4. An expert can be believed, by virtue of credential, training, education, profession, publication or experience, to have special knowledge of a subject beyond that of the average person, sufficient that others may officially (and legally) rely upon the individual's opinion.

The commonality of these definitions is the word "experience."

What is your experience???

I was speaking to a group of young mortgage professionals about their expertise. They had recently graduated from college and were starting their first jobs in the mortgage industry as loan originators. We were creating talk tracks for them to use to set up meetings and workshops with real estate agents in their area. They were concerned about meeting with experienced real estate agents and having them ask questions they could not answer. The first thing I shared with them is that no one has all the answers! Simple! I've been doing this for over 20 years, and I don't know all the answers, and those experienced real estate agents don't know all the answers, either. Don't worry about what you don't know; focus on what you do know.

Let's say that again, in slow motion this time…

Don't worry about what you DON'T KNOW!

Focus on what you DO KNOW!

How slow did you really say it?

    Was it just a little slow?

        Was it a whole lotta slow?

Did you slow it down so much that you used your mouth to exaggerate each word so that you looked like you were in one of those slow-motion videos?

Ok, not really, but it is fun to do that, and when you do that, you remember it. Trust me, you will. If you come to one of my workshops, you are going to come up to me and you are going to say **Focus on What YOU DO KNOW!!!** Or at least I'm throwing down that challenge for you now.

Yeah, yeah, I'm moving on. Back to my story.

I was coaching my young mortgage professionals, who were freaking out because they don't know jack about the mortgage business. Who cares! There are a ton of people who know about the mortgage business! What these young professionals have for their "expertise" that these "experienced" real estate agents don't is their age.

Seriously!

By being young, they can tap into a HUGE market of first-time home buyers. They know about social media, they know how to use video, they know how to reach, market, educate and talk to an entire generation! They know about the concern of saving money; they know what it's like to have student loan debt.

Their lack of experience is really their strength!

It's their superpower.

It's their ninja shazam!

We went to work developing an entire strategy to leverage their "newness." We created a campaign and called it, #lovemyhome.

For these young mortgage professionals, their expertise was their age. Instead of having them go into a real estate office and say, "Hi, I'm new and know nothing, please send your buyers to me," we sent them into real estate offices armed with the #lovemyhome campaign that offered them an opportunity to be a part of the movement they were creating in their area to educate young professionals about the benefits of buying a home. We created educational material that covered topics like, how to save money for a down payment; how to buy a fixer upper; how to fire your landlord; the benefits of homeownership, etc. Our goal was to build edumarketing campaigns that would educate their generation about buying a home! Once we had created campaigns that they could relate to, they felt confident going into real estate offices and inviting them to their "cause." It also created an opportunity for real estate agents to take these young professionals under their wings and help mentor them in the process.

Remember;

> it's not what you **DON'T KNOW**,
>
> it's what you **DO KNOW**!

If you don't know all the answers, you can partner with someone who does and work on content together or you can hire someone. Most of you reading this have a TON of experience doing what you are doing, so it will be simple to do this.

Let's dive into **Step ONE**, which is discovering **YOUR** expertise!

Step One – Discover Your Expertise

If you are a financial planner, your expertise is helping people invest, save and increase their wealth.

If you are a mortgage professional, your expertise is helping people qualify for financing a home of their dreams.

If you are a chiropractor, you help people stay healthy and release their pain.

If you are a real estate agent, you help people buy the home of their dreams or create wealth with investment properties.

If you are an attorney, you help people start a business, or protect their business, or stay compliant, or whatever you may specialize in.

The first question you must answer may seem very simple to you, but it is important to lay the foundation of who you are, what you do and how your prospects and clients can quickly identify why they would do business with you. It begins by answering the following questions:

1. Who are you?
2. How do you help your customers?

I'll give you an example of how I answer these questions:

I am an edumarketing specialist. I help companies and individuals develop and implement training and marketing to position themselves as experts in their fields.

It is important for you to have a clear idea of what you do and how you help your customers so that you can quickly tell them why they might use you.

For some, this is called your UVP, or unique value proposition. Let's look at another example:

If you are a real estate agent, your answers might look like this:

I am a real estate agent. I match buyers and sellers and help them meet their homeownership goals.

You may only want to work as a listing agent, so you may only focus on getting listings, so you may only state the following:

I am a real estate agent. I help homeowners list their homes so that they can realize the highest return on their investments. I do this by helping them prepare and show their homes with little stress and worry.

You get the idea.

Take a few minutes and answer the questions for yourself:

1. Who are you?
2. How do you help your customers?

Activity

Who are You?

How Do You Help Your Customers?

Chapter 4

WHO ARE YOUR CUSTOMERS?

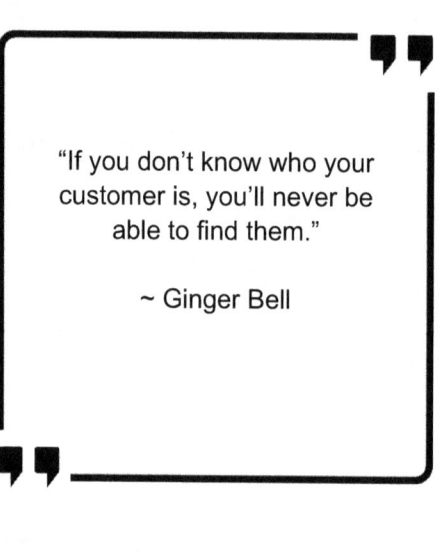

To identify what your customers want to know about, you must first understand who your customer is and what are their needs.

Fact: 86% of consumers say personalization plays a role in their purchasing.

Knowing your customers' needs and the questions they may have about making a purchase or doing business with you is the first step in creating your edumarketing plan.

Let's start with step one, which is identifying your ideal customer.

Step One: Identify Your Ideal Customer

Here is what that may look like if you are a real estate agent.

Ideal Customer(s):

1. Someone who wants to purchase a home.
2. Someone who wants to sell their home.

Pretty basic, huh? Now, we will get into more detail about their wants and needs and how to drill down on what they need to have to really qualify as a customer, but at the end of the day, if you are a real estate agent, your ideal customer wants to buy or sell a home.

How about if you are a mortgage originator? Who is your ideal customer?

Mortgage Originator Ideal Customer:

Someone who wants to finance a home.

How about if you are a financial planner?

Financial Planner Ideal Customer:

Someone who wants to invest.

And if you are a builder, who is your ideal customer?

Builder Ideal Customer:

Someone who wants a new home.

Remodeler?

Remodeler Ideal Customer:

Someone who wants their home to be more functional, nicer or updated.

This doesn't have to be complicated. Your ideal customer has a need, and you have a solution.

So, take a moment and identify your ideal customer.

Activity

Identify Your Ideal Customers
_____
_____
_____
_____

You may have several ideal customers depending on your product or service offerings. Clearly defining your ideal customer will help you focus your edumarketing efforts.

Next, complete the second step and drill down on the needs of your ideal customer. Nothing is more powerful in marketing than a direct message to a prospective customer, client or patient. Absolutely nothing. They see in your message that you understand them and speak to their specific questions, problems and needs. "Micro-targeting" to specific segments means you can leverage your information and speak directly to them. Let's look at how to identify your "micro-targets" for specific segments of your ideal customer.

Step Two: Identify Specific Segments of Your Ideal Customer

If you are a real estate agent, your customer wants to buy a home. They may want to buy a home for their personal use or as an investment property. They may be looking to downsize or buy a condo. They may be looking for a second home for a vacation property. Each of these types of "segments" becomes an opportunity for you to create edumarketing campaigns focused on those particular segments. When you identify those segments, you can easily create content to help them answer their questions and in doing so, you become the expert.

If you take the time to identify your customer segments, you will have an easier time making a list of topics and content. Let's look at the real estate agent and their ideal customer segments.

Real Estate Agent Ideal Customer:

Someone looking to buy:

1. Their first home
2. A bigger home
3. A new home
4. A smaller home
5. A vacation property
6. An investment property
7. A multi-family home
8. A condo

Each of these specifics still relates back to their ideal customer—someone who wants to buy a home—but gets more specific about where that customer may be in the buying process and speaks specifically to their questions and needs.

Now, you make a list of some of your customer segments. Take some time to complete your list.

Activity

Identify Your Customer Segments

Chapter 5

**WHAT PROBLEMS DO YOU SOLVE?
WHAT QUESTIONS DO YOU ANSWER?**

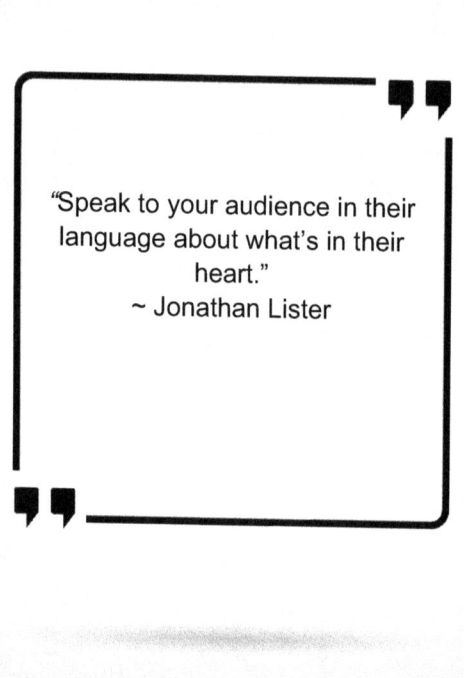

Your customers have problems. Your customers have questions and guess what? You have solutions and answers for them. Using that information is the key to building your edumarketing plan. First, you need to understand what those problems and questions are, so let's get started.

To drill down on your expertise, answer the following questions:

1. What problems do you solve?

2. What questions do you answer?

Let's answer these questions using our young mortgage professionals as examples:

Question

What problems do you solve?

Answer

The problems that these young mortgage professionals solve is the ability to relate to a young demographic. The average age of real estate agents is over 55. The industry is also seeing delays in young professionals buying homes because of perception of what happened with the mortgage meltdown and high student debt. By being able to relate to this demographic market, they can build around their expertise.

Take a moment to answer what problems you solve for your customers.

Activity

Identify the Problems You Solve for Your Customers

Question

What questions do you answer?

Answer

Although our young mortgage professionals were new to the mortgage industry, they were not new to the questions and discussions that their friends were having about the next step in their lives, which is buying a home. All they need to do is hold a few discussion groups at the local brewery and find out what young professionals want to know about buying a home. If you don't know what questions your prospects are asking, then I recommend writing down the questions your customers are asking on a regular basis. If you still don't know, then send out a survey and ask them what they want to know about. You can also search online to find out what others who do what you do are talking about. Search for trends of what people are searching for online. Join discussion groups. There are many things you can edumarket on, and if you still cannot figure it out, contact us. We can help you identify your expertise and build an entire edumarketing campaign around it. Seriously, that's what we do!

Take a moment and write down the questions you answer for your customers.

Activity

Identify the Questions You Answer for Your Customers

Let's move on to the third question.

Question

What specialized solutions do you offer?

Answer

Normally when I am working with a customer in the mortgage industry, I would build an edumarketing campaign around their loan programs, but in the case of our young mortgage professionals, they didn't know enough about loan programs to feel comfortable talking about them, and that is why we created the campaign #lovemyhome.

We started by having them talk about the challenges young professionals face when planning for their first home purchase. This allows them to start building relationships with real estate agents and potential homeowners while learning about various loan programs. We put them into a mentoring group who were able to help them get up to speed quickly and allow them to have fun educating and building the market in their area.

I don't want you to start focusing on your product or what you do at this point. What I want you to think about is what SOLUTIONS you offer. Solutions are answers. Solutions are answers to questions your prospects have when they think of you. If you are a chiropractor, you provide hands-on spinal manipulation and other alternative treatments to help align the body's musculoskeletal structure. But you may also provide nutrition guidance or life coaching. Look at all the solutions you offer, not just the services you provide.

Once you have identified your expertise, you can move on to the next step, which is to identify what your customers want to know.

Activity

Identify the Specialized Solutions You Offer Your Customers

Chapter 6

WHAT PRODUCTS/SERVICES DO YOU OFFER?

"Marketing is no longer about the stuff you make, but about the stories you tell."

~ Seth Godin

Identify Your Product Offerings

Now that you've identified your ideal customers and customer segments, take a moment and identify your product offerings.

Edumarketing can bring your business and customers together. Understanding those two elements is the foundation of identifying what your customers want to know about. The first steps to take are to answer the following questions:

1. Who is your ideal customer? (Remember your customer segments)
2. What do they want?
3. What are your products and services?
4. What problems do your products and services solve for your customers?
5. What questions do you answer?

Who Is Your Ideal Customer?

We went through the exercise of identifying your ideal customer and your customer segments in previous chapters. It is important to remember to focus on specifics of your customer segments and drill down on how each of your products or services will help them. You may have crossed over into many of your customer segments, and that is fine, but as you go through this exercise, focus on your customer segments and your products and services and how it can help each of your customer segments.

Activity

Identify Your Customers

What Do Your Customers Want?

Seems like a pretty basic question, doesn't it? But, do you really know what they want? Can you give them what they want with your products and services? How do you know? Being able to know what your customers want is an important step to creating your edumarketing plan. Remember, each of your customer segments may have a different want. Let's look at an example for our mortgage originator.

Ideal Customer: First-time Home Buyer

What Do They Want: To be able to buy an affordable home in their budget.

It's not rocket science, but it is important to know what your customer wants.

Take a moment and complete what your customer(s) want. You don't have to do this right now for all your customer segments, but eventually you will want to complete this activity.

Activity

Identify What Your Customers Want

Now, we will move on to identifying your products and services. Each of your products and services will provide a solution for your customers. It will be an answer to their wants, needs and questions. An essential part of getting your edumarketing plan right is understanding your audience.

Let's look at the products and services our mortgage originator offers:

1. Low down payment programs
2. VA home loan programs
3. Condo loan programs
4. Renovation home loan programs
5. Investment property loan programs

You should already have a good understanding of the products and services your business provides. You should also know who your ideal customer is and what they need. Let's begin with the basics. What products and services do you offer? Take a few minutes to list out the products and services you currently offer. We will line everything up soon, so don't worry about your customer segments yet, just complete the activity listing the products and services you currently offer.

Activity

What Products and Services Do You Offer?

Your next step is to identify what problems these products and services solve for your customers.

We will use our mortgage originators as an example.

Products/Services	Problem It Solves for Your Customer
Low down payment programs	Allows First-time Home Buyer to buy a home with little money down.
VA home loan programs	Provides 100% financing for qualified veterans.
Condo loan programs	Provides financing for someone who wants to buy a condo.
Renovation home loan programs	Wraps renovation costs into a home loan so homeowner can get updates to a home they are purchasing or refinancing.
Investment property loan programs	Gives flexible financing opportunities for individuals wanting to do fix and flips or buy investment properties.

By taking each of your products or services and breaking each down into problems and solutions, you can come up with several ideas to help you start creating your edumarketing.

Now, take each product and/or service you listed previously and answer the problems that each solves for your customers.

Activity

Products/Services	Problems It Solves for Your Customers

Identify Your Customers' Questions

Your next step is to identify what your prospects and/or customers want to "learn" about. What questions do they have that you can answer for them? Let's look at an example for a mortgage originator, who has a first-time homebuyer as one of their segmented customer categories:

Ideal Customer	What Do They Want?	Product/Service	Problem it Solves	What Are Their Questions?
First-time homebuyer	To buy a house	Down payment assistance programs	Allows them to get into a home with little or no money down. Can buy a home faster.	How much do they need to have in savings? How much do they need to earn? How much money can they borrow? What is the process?

Let's look at another example. Say you are a financial planner and one of your customer segments is someone between the age of 30 and 40 years old who wants to set up their long-term financial plan. They want to save for retirement, and they want to save for their children's college education.

We begin by identifying the ideal customer and what they want and match it to your product offering, then we write out what questions they may have that you can build your edumarketing plan around.

Let's look at the example for our financial planner:

Ideal Customer	What Do They Want?	Product/Service	Problem it Solves	What Are Their Questions?
30-40-year-old looking to build a plan for retirement and education for their children.	A long-term plan to allow them to retire at a certain age and provide college education for their children.	IRAs 403b 401k	Long-term savings and investment planning.	How much do I need to get started? How will I know if I'm on track? Am I making the right investments? What are the tax benefits?

Now, it's your turn to complete this exercise for your ideal customer segments. Be sure to do this for each ideal customer category because you will want to build a specific edumarketing plan for each customer segment. This is critical because it will allow you to tailor specific education to each segment's needs. Why is this important? Because if you are an investor looking to buy a rental property, you already know how to save for your first home, so watching a video on how to save for a down payment may not be of interest to you. You would be much more interested in watching a video on finding the right tenant or what properties are best for rental income. The more specific you can get to answering the questions your ideal customers have, the more they will see you as a trusted expert. Remember, you don't have to be everything to everyone. Identifying your ideal customer will allow you to get specific and create that expert niche you are looking for. Take a moment and complete this task for each of your customer segments you identified earlier. Match them with your product offering, and then write out what questions they may have. Complete this exercise for all segments you identified for your ideal customer.

Activity

Ideal Customer	What Do They Want?	Product/Service	Problem It Solves	What Are Their Questions?

Let's recap the steps to finding out what your customers want to learn about.

1. Who is your ideal customer? (Remember your customer segments)
2. What do they want?
3. What are your products and services?
4. What problems do your products and services solve for your customers?
5. What questions do you answer?

Before we move on to creating your edumarketing plan, I want to take a moment to talk about creating edumarketing campaigns for your referral partners and existing customers. Let's not forget that your customers are not just those who buy one time from you. Your existing customers can and may be a great referral source for you. They can also be repeat customers, so you will want to look at building long-term edumarketing plans for your existing customers to continue to buy from you in the future. By being that resource and trusted expert, you will build customers for life!

Identifying Referral Partners

Yes, you guessed it—another activity for you to complete. Every business, no matter who you are and the product or service you offer, has potential partners. If you are a mortgage originator, your partners are real estate agents. Why? Because their customers want to buy homes, and you offer the financing programs to help their customers buy homes. If you are a financial planner, your partners could be mortgage originators, insurance agents, real estate agents, attorneys, etc. If you are a chiropractor, your partners could be nutritionists, trainers, insurance agents, or attorneys. The reality is that everyone has a database of customers, and if there are potential questions that someone's customers have that you could answer for them, then you can build an edumarketing campaign for or around them that can help you build your business.

The first step is to identify your possible referral partners.

Let's look at the example of the mortgage originator. Their potential referral partner list might look like this:

Potential Referral Partners for a Mortgage Originator

REAL ESTATE AGENT
ATTORNEY
INSURANCE AGENT
FINANCIAL PLANNER
HOME REMODELER
HOME DECORATOR
LIFE COACH (Yes, really, a life coach is helping someone plan their life, why not include home ownership in that plan?)

Now, your turn! Go ahead and list your potential referral partners.

Activity

Who Are Your Potential Referral Partners?

_____
_____
_____
_____
_____
_____
_____

Next, identify your referral partners and their customers. In other words, find a match.

Let's look at our mortgage originator example:

Referral Partners and Ideal Customer Match

REFERRAL PARTNER	IDEAL CUSTOMER
REAL ESTATE AGENT	Someone who wants to buy a home and finance it. Could be a first-time home buyer. Could be investment properties. Could be condos. Could be a second home.
ATTORNEY	Someone who wants to buy a home. This could be an attorney who is working with someone on planning an estate. You could provide education on creating a trust and how to purchase properties inside a trust. You could partner with an elder attorney and provide information on reverse mortgages for retirement planning.
INSURANCE AGENT	Someone who owns a property and wants to make sure they are properly insured.
FINANCIAL PLANNER	Someone who wants to build wealth, plan for their future and include investment properties in their portfolio.
HOME REMODELER	Someone who wants to stay in their home and fix it up. Great renovation

	products allow you to market to their database and bring them business too.
HOME DECORATOR	Someone who wants to stay in their home and fix it up. This is a great partner to work with along with a home remodeler.
LIFE COACH	Someone who wants to build their life plan. This is an outside-the-box referral partner, but they have customers and are having conversations with them about life planning, so why not bring you in as an expert to talk about real estate investment planning as a part of their life plan?

Once you have identified your potential partners and ideal customers, you can match them up with your product offerings and questions that you completed earlier. Some of the edumarketing campaigns you will develop can be used to send to your referral partners or repurposed to answer specifics that relate to their customers' questions. You will also want to utilize your partners in your edumarketing campaigns because the information they provide to their customers could also be of value to your customers. Developing partnerships is vital to an effective edumarketing plan.

Let's move on and briefly talk about your existing customers and what you can do for them to build an edumarketing referral campaign.

You have a database of existing customers who trusted you one time and would probably trust you again, but you forget to either continue to provide information to them or think they won't refer people to you. Getting referrals is one of the most powerful strategies for gaining access to new clients.

When you are referred to someone, there is increased credibility and trust, which accelerates the sales process.

The reality is that an introduction from a happy client is a heck of a lot easier than making multiple phone calls, sending emails or attending networking events.

The unfortunate reality is that most people don't ask for the referral. Why? Usually people don't ask for a referral because they see it as being too forward. That is why building an edumarketing campaign to continue to stay in touch with your existing customers, answer their long-term questions and provide information that they can share with their friends and acquaintances will allow you to discreetly stay in touch with your customers, continue to build value, provide information and be the expert.

Part of your edumarketing plan will include creating an edumarketing plan for your existing customers.

We've covered a lot in this chapter. Figuring out what questions your customers have will help you identify the information you are going to create for them. If you don't readily know your customers' questions or want to find out from them directly, consider sending out a questionnaire to your customers to find out what information you can provide them about your product or service.

You may also consider conducting an introductory webinar that covers your services and asking what questions they have. Services like SurveyMonkey have easy-to-use templates to create simple surveys you can create to ask your prospects what they want to know.

So far, we have covered why you should use edumarketing in your business, how to identify your expertise and what your customers want to know. Now, let's move on to discussing creating a content strategy.

Chapter 7

CREATE A CONTENT STRATEGY

"Success is the sum of small efforts, repeated day in and day out."

~ Robert Collier

In chapter two, we covered a quick start plan to creating content. The sooner you start adding edumarketing to your marketing plan, the better; but to create a long-term plan, you will want to create edumarketing based on your customer segments. Without a content strategy for your edumarketing, you are forced to educate and market to every customer the same way. You might as well put your message on a billboard or go door-to-door with this type of strategy. Instead, I am going to show you how to build a content strategy so that you can educate and market specifically to a particular customer and provide them with information based on their needs, wants, desires and questions. This will allow you to build trust with your prospects and leads and create long-term fans with your partners and customers.

With a complete content strategy, you will be able to tailor your message more closely to your customers' unique needs, keep them engaged with you and your expertise, and nurture them down the sales funnel.

Knowing your customers' interests, problems and questions is the key to building your content strategy. Your content strategy should act like a speedometer for where your customers are going, when, how fast and what information you can give them. When used correctly, a content strategy will help to specifically market, educate, build trust and make sales.

In this chapter, we are going to focus on your ideal customer and your customer segments; but remember, you also have existing customers and partners you can implement the same process with.

Right now, we are going to go back to what we covered in Chapter Four where we identified your ideal customer segments. These segments make up what we call a target market. Within each target market you have a specific "type" or segmented customer. These segmented customers each have questions that you can answer.

Begin with step one, which is to identify your customer segments. We did this earlier, so you can use your previous answers here and expand on them.

You can download the **Edumarketer Planning Workbook** at:

www.theedumarketer.com using the code **edumarketer**.

You may recall the example of our real estate agent. Their ideal customers were people who wanted to buy or sell a home. Their ideal customers look like this:

- Someone who wants to buy a home.
- Someone who wants to sell a home.

Now, It's your turn. Complete step one and identify your ideal customer(s). If you completed this earlier, you can copy your answers below. This is important because you will build edumarketing campaigns that speak to your overall ideal customer and for many, this is a great place to start. However, the most effective edumarketing campaigns are those that are targeted to your customer segments. You are educating based on their needs, wants, desires, interests and questions. We will get into more detail on that later.

Step One:

Identify Your Ideal Customer

You've got your ideal customer, but, remember, we then broke these down into more specific segments based on each customer's needs and interests.

For this part of your planning we are going to get very specific about your segmented customers so you can create your edumarketing plan for them.

For example, if you are a real estate agent, you may want to create one edumarketing plan for first-time homebuyers and one for investors. You will not be creating the same content for these two segments. The more specific you can be to the education you are creating for your segmented customers, the more likely you are to build trust as the expert.

Our real estate agent's segmented target customer list for someone who wants to buy a home could look like this:

- Someone who wants to buy their first home.
- Someone who wants to buy their second home.
- Someone who wants to buy a bigger home.
- Someone who wants to buy a smaller home.
- Someone who wants to buy an investment property.

Your turn. Complete your customer segments.

Step Two:

Identify Your Customer Segments

Step Three:

Your next step is to identify the questions each customer segment may have. What do they want to know? What can you teach them? What questions are they asking, and what information can you provide them that will answer their questions and position you as the expert to trust and buy from?

Let's pull it all together and look at our real estate agent example:

Customer Segments	What Do They Want?	What Are Their Questions?
First-time homebuyer	To buy their first home	Where do I start?
		What are the benefits to owning vs. renting?
		How much money do I need to have in savings?
		How do I pick a real estate agent?
		When is the right time to buy?
		What's the best neighborhood?
		What is the process of buying a home?
Investor	Buy an investment property	How much do I need for a down payment?
		What type of home should I buy?
		What are the benefits to owning an investment property?
		Should I buy a fixer-upper?
		Do I need to be handy?
		What loan programs are available?

Your turn. Complete your list.

Customer Segments	What Do They Want?	What Are Their Questions?

Step Four:

Ok, hard part's done. Now we get to have FUN! Let's look at what edumarketing campaigns you can build to answer your customers' questions.

Step four is to identify your edumarketing campaigns. Let's see what that looks like for our real estate agent.

Customer Segments	What Do They Want?	What Are Their Questions?	Identify Your Edumarketing Campaigns
First-time homebuyer	To buy their first home	Where do I start? What are the benefits to owning vs. renting? How much money do I need to have in savings? How do I pick a real estate agent? When is the right time to buy? What's the best neighborhood? What is the process of buying a home?	Are you ready to buy your first home? A complete guide to getting ready to purchase your first home. - Where do you start? - What are the benefits? - How much do you need to have in savings? - How to pick your agent - Looking for a home - Making an offer - The home buying process - Closing on your home - Selecting a mortgage specialist
Investor	Buy an investment property	How much do I need for a down payment? What type of home should I buy? What are the benefits to owning an investment property? Should I buy a fixer-upper?	Top tips for buying a rental property. - What types of properties are good for investments? - How much money will you need for a down payment? - Should you do a fixer-upper? - How much rent can you charge? - Finding the right

| | | Do I need to be handy?

What loan programs are available? | | loan program |

Your turn. Take your time and complete your edumarketing campaigns for each specific ideal customer. Remember, you can download the complete Edumarketer Planning Guide at **www.theedumarketer.com** using the code **edumarketer.**

Customer Segments	What Do They Want?	What Are Their Questions?	Identify Your Edumarketing Campaigns

Now that you have identified your customer segments, their wants, questions and your edumarketing campaigns, it's time to develop your edumarketing plan, and we will do that by writing your topics.

Chapter 8

UNDERSTANDING THE PROCESS

"Good marketing makes the company look smart. Great marketing makes the customer feel smart."

~ Joe Chernov

The content you develop will help to educate your prospects and customers and answer questions they have regarding your products and/or services. When developing educational content, it is important to first understand the steps in the education process.

There are six basic steps in the education process.

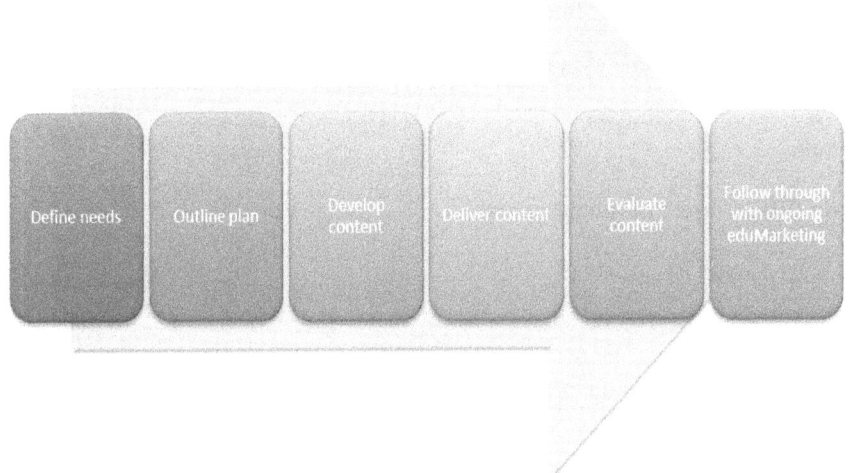

The edumarketing process is repetitive. Every time you begin to consider edumarketing, you must evaluate who your customer is, what are their questions and what is your delivery channel. We can only see if your edumarketing was effective if we follow up afterward. We will discuss this further in an upcoming chapter.

Before your eyes start glazing over, follow this plan.

Step One: Write Down General Topics

If you are a mortgage professional, your topic may be saving money for a down payment, or benefits of buying vs. renting. Write out several topics. The idea is to discover what topics you want to start building your content around. You don't have to do anything more than just write down topics. Take a few minutes and write down at least five topics. You may come up with more than five topics. Just focus on your topics. Think about what questions your prospects or clients have. In previous chapters, you identified your ideal customer, customer segments and specific needs; go back and look at your list. This will help you list out various topics. Topics are very general and can come from your segmented customer list.

We will use our real estate agent as our example again.

Topics
1. Buying Investment Properties
2. Buying Your First Home
3. Buying a Condo
4. Buying a Fix and Flip Home

Each of these topics relates to a segmented customer who is searching for information. By selecting information on a topic, they may be searching on, you place yourself in front of them in their search.

Now it's your turn. Identify a few topics you can use to develop your edumarketing content.

Activity

Write Out Several Potential Topics

Step Two: Write the Content Headlines You Can Build For Each Topic

Now that you have identified your topics, let's drill down on questions, answers and information your prospects and customers are looking for. Content headlines become the headline that will show up in online searches or articles you write. The content headline is what will grab the attention of the viewer.

Let's look at an example of the topic: Investment Properties for our Real Estate Agent:

Topic: Buying Investment Properties

Content Headlines
Finding the right investment property
Hot properties
How to pick the perfect investment property
Top 5 features of a profitable rental property
Top tips for buying your first rental property
5 Things to consider when buying an investment property

Take a few minutes with one of your topics and write out some ideas for content you can develop. If you are looking for ideas, go to Google and type in your topic. Google will tell you what people are asking for, and you can begin developing your content ideas from there.

Activity

Write Your Topic Headlines

Step Three: Determine Your Delivery Channels

Once you have identified your topics and content headlines, you can move on to determining your delivery channel(s) so that you can develop your content deliverables. In the next chapter, we will discuss the various delivery channels you can use for your edumarketing plan. You will eventually want to use multiple channels for the highest results. Video is often one of the first delivery channels selected; however, you may want to start by developing content for your blog and build into developing videos. You may want to create an online learning portal that encompasses both video and written content. Whatever you choose to do is fine, but you will want to determine your best delivery methods and just get started on them; otherwise, your edumarketing plan can become too complicated and you will lose traction. Best practice is to choose one delivery channel and create your content and deliverables for that channel and then repurpose from there.

Let's look at our real estate agent as an example.

Identify a Topic

In our example, we will have our real estate agent create content on the topic of **"Investment Properties"**.

Write Content Headline

From this topic we will write the content headline: **"Five Things to Consider When Buying an Investment Property"**

Determine Delivery Channels

In this example, the real estate agent has identified creating content for a blog article and a YouTube video. They are going to create two deliverables for their headline by writing a blog article and using the content they develop in the article to also create a video to post on YouTube. You can often use multiple delivery channels for the same topic and link the two together.

Let's look at an example:

Topic	Content Headline	Delivery Channel(s)
Investment Properties	5 Things to consider when buying an investment property	Blog Article YouTube Video

Take a moment and do this for one of your content headlines.

Topic	Content Headline	Delivery Channel(s)

Now that you have determined your topics, content headlines and delivery channel(s), you can move on to developing your deliverables.

Chapter 9

**HEADLINES COUNT!
CREATING AWESOME HEADLINES**

"If your stories are all about your products and services, that's not storytelling. It's a brochure. Give yourself permission to make your story bigger."

~ Jay Baer

By the end of 2019, it is projected that we can expect to see 2.9 billion worldwide email users[iv] (this translates to more than one-third of the global population). According to Statista, almost 320 billion daily emails will be sent in 2021[v]. On average, Facebook generates more than 8 billion daily views of video content, 85% of Facebook users consume video content with the sound off, and 78% of American consumers say they've discovered products on Facebook. With more and more information coming at us each and every day, having a headline that stands out is critical to getting your information viewed.

Your headline is what will attract the viewer to read, watch or find out more information about what you are talking about. A headline is simply a sentence which in essence describes your information in short form. Many people only read headlines when skimming through email, social media and searches, so having a headline that will grab their attention to go on and find out more information is important. Your headline must stand out on its own.

For most of you reading, you don't have a background in copywriting or marketing, so knowing how to write a catchy headline may cause you to freeze up. Don't worry, there are some simple rules to follow when writing catchy headlines that will make it easier for you to come up with your catchy headlines. Let's first look at the basics of headline writing; I like to think of it as writing a menu. When you go to a new restaurant, what's the first thing you do when you open the menu? You glance through and look at what? The titles of each entrée. You look for what catches your eye or stomach at the time. A simple menu item titled "Hamburger" is not nearly as enticing as "Savory BBQ Bacon Cheeseburger." "PBJ" is not as exciting as "Childhood Worthy Peanut Butter and Grape Jelly Smothered Sandwich." You get the idea. Details count, and people want to be enticed, so entice them. Write your headlines like you are writing a menu. Let's start with the basics:

Step One

Start by writing a simple how-to headline. For example:

- How to Find the Perfect Home
- How to Feel Less Pain
- How to Save Money For Retirement

Step Two

Use your how-headline and add words that stand out and tell the viewer the benefits of reading your post, attending your workshop or watching your video. Let's look at our examples:

How To Find The Perfect Home

- Three Easy Steps To Finding The Perfect Home
- The 10 Most Common Mistakes People Make When Looking For The Perfect Home
- How To Find The Perfect Home Online
- Create The Perfect Home Of Your Dreams
- Top 5 Mistakes People Make When Searching For Their Perfect Home

How To Feel Less Pain

- Three Easy Steps To Feeling Less Pain
- The 10 Most Common Mistakes People Make When Searching For The Perfect Pain Relief
- How To Start Feeling Better And Feel Less Pain
- Create A Better Day, And Feel Less Pain
- Top 5 Mistakes People Make When Looking For Pain Relief

How To Save Money For Retirement

- Three Easy Steps To Saving Money For Retirement
- The 10 Most Common Mistakes People Make When Saving For Retirement
- How To Start Saving For Retirement Today
- Don't Think You'll Ever Retire? Create A Retirement Plan Today For Your Dreams Of Tomorrow
- Top 5 Mistakes People Make When Saving Money For Retirement

You can see how each headline uses descriptive words to attract the viewer to want more. Of course, you need to make sure that the information in the headline is actually what you are providing the reader. If you follow the steps I've provided, you will start by identifying your topic, then writing your headline and finally developing your content. The headline will give you an idea of what to write, vlog and talk about.

Let's look at some other top ideas for headlines that work.

1. Lists

Think David Letterman. He put Top Ten lists on the map, and people love them. Lists can provide useful bits of information that are simple to develop. Don't go beyond 10 on any list. Three to five is usually best. If you have more than that, you have a series and can move your information into segmented content. Here are some examples:

- 10 Smart Ways To Finance Your Rental Home
- 6 Little Steps to Saving for Retirement
- 5 Ways to Improve Your Bone Health

2. **How-to Guides**

People are searching online to learn "how to" do something. A how-to article, guide, list, or video is meant to teach, and that is exactly what edumarketing is all about. It is easy to create how-to content if you base it around your customers' questions and then provide the answer. Remember to identify the most common questions your customers have and then you can quickly develop your how-to headline and content. Below are some examples:

- The new revolutionary how-to guide to selling your home
- How to buy your first investment property
- How to fix and flip an ugly house

3. **Resources**

Resource content is usually a lengthier guide that you can reference in a video or article. Resources can be used to entice your audience to give you their email in exchange for the guide or can be used in a series where one leads to the other. Resources can be guides, lists, how-to's, etc. Let's look at some examples:

- The Beginner's Guide to Buying Rental Properties
- Ultimate Guide to Saving for Retirement in Your 20's!
- Drug-Free and Pain-Free, the Ultimate Guide to Living Life without Pain

4. **Questions**

Questions pose the ultimate answer to what your viewers are searching for. It defines something that they can relate to. Questions are meant to not just ask the question but provide the answer. Questions can be extremely valuable in your edumarketing plan. Let's see some examples:

- How Long Should It Take to Sell My House?
- If I'm So Smart, Why Aren't I Ready for Retirement Yet?
- How Do I Start Saving to Buy a Home?

It is easier than you think to create headlines. Remember these final tips to creating headlines that convert:

1. Make it personal. Remember to write to your "micro" or target customers. Write in a voice that speaks directly to them as if you are having a one-on-one conversation with them.
2. Make it powerful. Your headline will make the viewer want to move on and read the first sentence or view your video.
3. Make it useful. Your headline should speak to answering a question or solving a problem. It should relate to what your customer is seeking and position you as the one and only who can help them solve their problem or answer their question.
4. Make it memorable. People do not remember topics, headlines or information that is dull and boring. Think about how you are going to tell them about that delicious item that is on your menu. Make them want it, crave it and desire it.

Your headlines will build a memorable experience for your audience and position you as the expert you are. Remember to take the time to write a headline that will capture attention, answer questions and make them want more.

Chapter 10

DEVELOP YOUR CONTENT

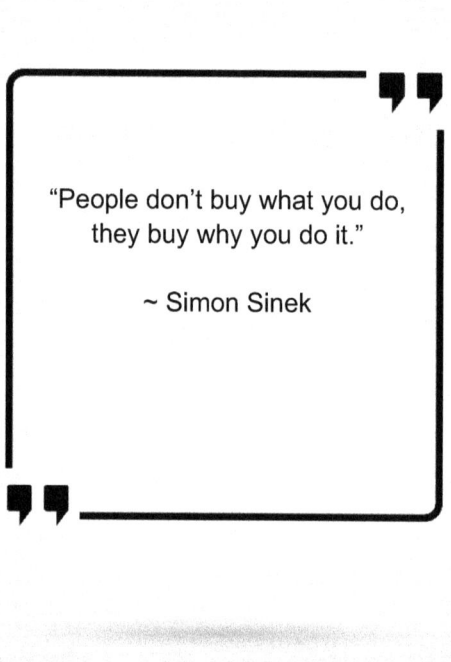

There are many options for creating edumarketing content. You may write a blog post or create a video for your YouTube channel. You may create a check list to use to capture leads on your website or an online course for your eLearning site. You may create a PowerPoint slide deck to use in a webinar or you may use the topic as a discussion in your podcast. You can develop several different deliverables for each topic. The most important thing to remember is to take your topic and write your content, then develop your deliverables. Yes, this process does take a bit of up-front work, but once you have done it, you will be thankful because you will end up with a strong list of topics that will support you in your marketing and business goals and position you as the expert in your field.

As you go about creating content, you will want to create a content hub to organize your content for your various distribution channels like blog posts, videos, social media, presentations, eLearning, eBooks, etc. If your topic is long, then you may want to create one cohesive paper and/or presentation. Although the content you create may be divided into multiple deliverables, creating one single presentation or whitepaper gives you more control and will help you:

- Divide your broader topics into smaller topics
- Develop multiple content for various distributions channels
- Build FAQ's

Let's look at our real estate agent again and see what they will develop for their deliverables.

Topic	Content Headline	Delivery Channel	Deliverables
Buying Investment Properties	5 Things to consider when buying an investment property	Blog Article	Begin by writing an article in Word on the 5 things to consider when buying a home. Will use links to sites like Forbes to back up information. Once article is written, post onto blog. Will be sure to add Google search words like investment properties, buying a rental home, etc.
		YouTube Video	Develop the video from the article. Will record the video from the top 5 points in the article. The video will only be one minute long and can reference back to the article for more information. Video will be highlights.
		Downloadable Checklist	Create a checklist for buying an investment property. Create a link to downloadable in CRM and capture email to build marketing database.

By taking the content headline, you can create the content, in this case an article, then from the article, you can develop other deliverables. With a plan like this, you can take one content headline and develop a blog article, short YouTube video, or a webinar where you take the content in the article and create a slide deck where you talk about each of the five points. You could then develop a breakout session to hold at a local association chapter meeting. There are so many options, but the key is to begin by identifying the topic, getting creative with your content headlines, determining your delivery channels and then creating the content. In our next chapter, we will move on to discussing how to market your edumarketing content.

Now, it's your turn. Take one of your topics, write out the headline, determine your delivery channels and then write out your plan for creating your deliverables.

Activity

Topic	Content Headline	Delivery Channel	Deliverables

Chapter 11

CREATING MODULAR AND RUBBER BAND CONTENT

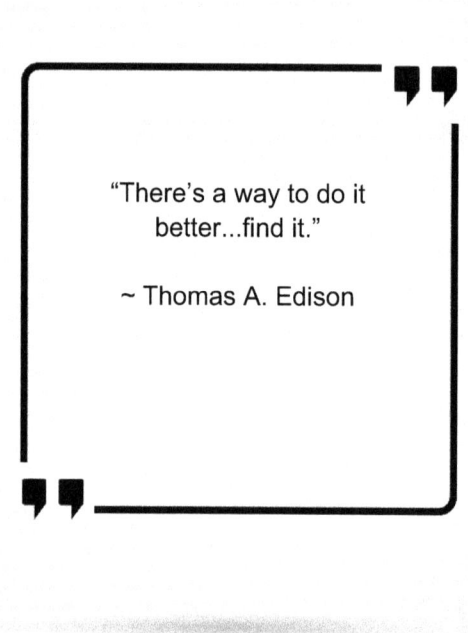

For most of my adult life, I have been creating corporate education programs. What started as live training and workshops in the early 1990s moved to online education in the early 2000s due to economic challenges. WebEx and Gotowebinar created the ability to train hundreds while sitting in front of your computer, and although most people hate online education, it has become a common method for training within most organizations. When you set out to develop a course, you have to think in terms of modules. The same holds true for your edumarketing content. If you create your content using a modular or what I call "rubber band" approach, you will be able to build a long-term following of prospects, customers and partners who will always look to you for answers to their questions and solutions to their problems. Modular content is the key to every successful edumarketing program. So, what does it mean?

Module Defined

A module is a unit, chapter, topic or segment of a larger topic. If you take one big topic and "chunk" it into smaller "snackable" pieces, you can create smaller content pieces. Most people's attention spans and communication habits are getting shorter and smaller. We need to deliver content to them in smaller "bite size" pieces. If you have an answer to a question but it's going to take 15 minutes to explain it to someone, then you want to break those steps into smaller 2 to 3-minute segments. Think *Sesame Street*. Short, quick bits of information that gives content that is relevant and to the point. You can still build information and lead the audience to more information if they want more. Create your overall content and pull from that content into smaller segments of content.

How Do I Start Building Modules?

Begin with the basics, which is understanding what your potential clients, customers and partners want to know about. Think about what questions and problems they have and what specialized services you offer. Segment your customers and then start building topics from your list of questions, problems, products and services and customer segments. Your topics can each become a larger piece of content that you can create smaller modules from. Take, for example, a loan originator. Their customer segment may be First-time Homebuyers. The larger topic is "How to Buy Your First Home." With this one topic, you can write an overall plan that may look like this:

1. The Benefits Of Homeownership
2. How To Move From Renting To Buying
3. How Much Do You Need To Save For Your First Home?
4. What Kind Of Loan Programs Are Available For First-Time Homebuyers?
5. When Is The Right Time To Buy A Home?
6. How Your Debt Plays A Part In Qualifying

Although the overall topic is "How to Buy Your First Home," each topic becomes a snippet and offers content related to buying your first home. When you create a broader topic, you are able to quickly break out smaller snippets of content. As you are planning your edumarketing, start with your topic, create your smaller segments or snippets and then write or record your content at one time. You can still create your content as one entire piece and then divide your content into smaller segments or snippets to be delivered over a period of time. Once developed, you can then take your content and divide it into smaller segments for distribution to your various delivery channels.

What is "Rubber Band" Content?

Rubber bands are used to hold things together. The key to their success is their ability to expand based upon what they are being used for. Same holds true for your content. Take your content and expand on it, build on it, and create additional content from it. If you are doing a video on buying a home, create a checklist that covers what they need to do to save money for a home, locate a home, find a realtor, or buy a second home. Each of your topics can expand into other content. Plan on creating additional content or takeaways from your topics and make your content live longer. Look to various distribution channels for ways you can "rubber band" your content.

Build Big – Break Small

The key to long-term content creation is to start with the bigger picture and then chunk that content into smaller snippets for distribution. Include additional takeaways that build on the content and allow the audience to research, find out more and continue to connect with you in your brand and expertise. Be sure to engage and include additional delivery channels.

Chapter 12

SELECTING YOUR DELIVERY CHANNELS

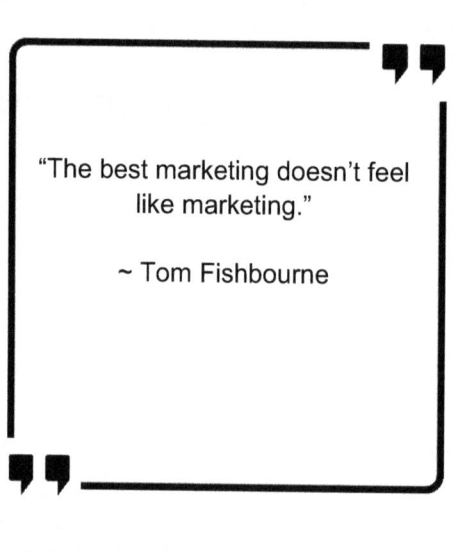

The best edumarketing plan includes a variety of distribution channels, including:

- Blogs on your website and partner websites
- Micro-blogs such as LinkedIn, Twitter, Facebook and Instagram
- Online videos
- Live workshops
- Podcasts
- Webinars
- Online education
- Radio shows
- Magazine and newspaper articles
- Whitepapers
- E-books
- Print books

If used correctly, edumarketing will provide long-lasting referral customers who will look to you, and only you, as the resource.

There are a variety of channels you can use for educating. When creating your edumarketing plan, it is important to remember to include your potential and existing customers as well as your aligned partners in your strategy, as they can be a valuable referral resource for your business. Let's dive a bit deeper into some of the channels we listed.

Blogs, Whitepapers, Articles and Books

A common method used in marketing today is called content marketing. Content marketing includes blog articles, white papers, industry publication articles, e-books and studies. Content marketing does not have to be complicated or long. In fact, shorter articles are more effective. Today's best content marketing strategies involve shifting from trying to find the perfect way to explain an idea or process to instead thinking of each step along the journey and slowing publishing articles to take the reader through the process. Simply put; instead of having them drink from the firehouse, provide them with the water they need at that moment.

Email Marketing Campaigns

Email marketing is one of the most widely used forms of marketing and probably one of the most ineffective. Think about how many emails you receive each day. The likelihood of someone opening your email about your product and its benefits are slim. That is exactly why incorporating edumarketing into your email marketing campaigns is critical. Imagine having your prospects register to receive a weekly or monthly email from you that provides bits of information to help them in their quest for your service. I'm not talking about a lengthy email. I'm talking about one piece of information that educates them or provides them answers to the questions they have. It could be a 30-second video on how to search online for the perfect home.

It could be a list of top three investments they should include in their portfolio. It could be a link to a recent article on Forbes about saving for retirement. Whatever it is that you are emailing to your prospects, or ideal customers, it needs to be valuable, short and helpful to them.

Recently I have been helping clients create email courses. An email course is a series of emails that are sent out over a period of time. We create a landing page that is sent out via social media, subscription list or placed onto a website. The landing page has information that generates interest to common questions your ideal customers may have. You offer to give them a complimentary course to find out the answers to these questions. This edumarketing campaign is fantastic because it is not simply one downloadable report they register for on your website, that then goes to a bunch of spam emails you send them. Instead, the email course is just that—a course that provides them a little bit of information delivered to their email box over a period of time, usually 30 days. It is very specific and produces results with little effort. If you want to find out more about how to set up an email course campaign, you can go online to www.edumarketingagency.com and register for our **Email Course Design** training. It's a perfect example of how to set up your own email course, and guess what? We'll show you how to do it!

Online Videos

YouTube has revolutionized how people "learn." Videos are one of the most effective vehicles for storytelling, educating and connecting with people. According to Statista[vi], people spend over 1,000 minutes per month watching online videos. YouTube has become one of the leading search engines on the web. Including video in your edumarketing plan is an absolute must and YouTube is not the only channel to use. You should include videos in your social media posts, email marketing and most importantly your online e-learning portal, which we will detail later.

Presentations and Workshops

One of the most effective methods of positioning yourself as the expert includes speaking at the front of the room. Local chambers often host education for their members. Associations are another great venue for speaking about your expertise. You should not only look at associations in your field, either. Think about potential related fields that could benefit from your expertise and become a valuable referral partner. If you are not comfortable with speaking, then put together a panel of experts. If you are a real estate agent, you could speak on the advantages of homeownership for First-time Home Buyers by compiling a panel that includes a Mortgage Professional, Home Inspector and Insurance Agent. The panel would provide information in their field of expertise, and it would take the pressure off one person speaking the entire time.

Being the "expert" at the front of the room is one of the most powerful methods of edumarketing. Be sure to "socialize" your speaking so that others see that you *are* the expert. You can do this by having someone take photos of your event with you at the front of the room. You can post the photos on your social media, website and newsletters.

Virtual and Online Courses

You don't need to have a complicated software or product to bring value to your customers by including online education in your edumarketing plan. While customer training may be a necessity in some fields, for most it is a valuable way to use your expertise to attract new customers as well as build rapport with your existing customers. It can also provide an increase in your SEO. It isn't necessary to create extensive "how to" guides to have a powerful education plan. If you are an Insurance Agent, you could create a course on "How to Vacation Ready Your Home." It could include an info-graphic or a downloadable checklist.

Remember, people are looking for information online, so having an online university of courses that people can take for free is a valuable method to increase your database. There are a variety of technology products you can use to host your online education, including live webinars, recorded webinars, e-learning courses, white papers, checklists, videos and how-to guides.

Podcasts and Radio Shows

Podcasts have become a powerful marketing tool. A podcast is a set of digital audio files that are available on the internet for downloading. Podcasts have become the new talk radio on mobile devices. Including Podcasts in your edumarketing plan can help you reach new audiences. Podcasts allow you to develop a relationship with a listening audience. Podcasts are easy to create and highly engaging. It is also a great way to develop public speaking confidence. Radio shows and community cable shows are also a viable method for building a new market share.

Look to local stations and find out what their audience has for questions. You could do a segment on "Overcoming a Stiff Neck" if you are a Chiropractor or "Inexpensive Decorating Tips" if you are an Interior Designer. Education can be entertaining if done correctly and a valuable means to market your expertise.

Internet TV Channels

Internet TV shows are a relativity new media medium, but one that you may consider if your plans are to create an in-depth edumarketing plan.

Services like Strimm (pronounced: "stream") is what they call, "TV as a service." It is an additional edumarketing tool to consider using to promote yourself and your services. Strimm is a live TV creation, scheduling and broadcasting platform. Services like Strimm provide an easy-to-use tool to create linear TV channels with scheduled shows, which can broadcast in real time on your own website and on

Strimm. Really though, you don't need a fancy service like Strimm to start your own TV channel. You can create a video TV channel with your own Wordpress site or YouTube channel. The idea is to create a series of shows that provide content and build value. The idea is to educate and inform. If you are serious about starting your own channel, then you will want to plan out an entire series. I recommend creating a schedule for your recording, editing, posting, etc. Whatever means you decide on for creating your channel, the power of your own brand and station is well worth the time!

As you can see, there are a variety of distribution channels you can use to deliver your edumarketing campaigns. Each has the potential to position you as the expert in your field. I would not recommend starting out using every delivery method. Instead, pick one and create and launch a campaign in that channel, then move on to the next. Which one should you start with? That depends on you and your experience level. Video is one of the most effective, but it is also one of the hardest if you do not have editing skills. First pick for me would be email marketing. You probably already have a list of prospects that you could quickly tap into, and it is faster to build a quick email course to launch to that list. We will look at determining your delivery channels in the next few chapters. For now, we've laid the foundation of possible delivery channels for you to consider. Let's move on to developing your edumarketing plan!

Chapter 13

BUILDING YOUR EDUMARKETING PLAN

"Marketing is a contest for people's attention."

~ Seth Godin

You've identified your ideal customer, customer segments and interests, topics and content and created a few deliverables, now it's time to take what you've learned and create your entire edumarketing plan.

Having an edumarketing plan allows you to use content pathways that will help you apply a limited marketing budget more effectively.

Let's look at the primary elements of a good edumarketing plan:

Set Your Edumarketing Goals

Once you've decided to edumarket yourself and your business, you need to set realistic and measurable goals to achieve over the next 6-18 months. Giving yourself a specific time frame allows you to develop enough content and activities to measure results.

Remember, the goal of edumarketing is to build yourself as the expert and attract new customers. In the end, yes, you want an increase in sales; however, the primary goal to begin is to establish your expertise.

There are several ways to create your edumarketing goals. It is easiest to separate your goals into categories. For example:

1. Your 1 to 6-month goals
2. Your 6 to 12-month goals
3. Your 12 to 18-month goals

There are two primary goals for your edumarketing plan.

Develop Your Content

Your first goal is to develop content. For example, your first goal may be to identify six topics, one for each month. From those six topics, you will write an article, produce a video and create a downloadable.

Add Followers/Fan/Subscribers

Your second goal is to increase your "fans," or followers. You know, the people who look to you for your expertise. The more followers you have, the more likely they will look to you at some time because you have established yourself as the expert. Sometimes prospects will convert immediately, depending on where they are in the buying cycle. For others, it may take some time before they become your customer. Staying top of mind with your expertise will help them to look to you when they are ready to purchase your product or service.

Consequently, you will want to also measure the number of "fans" you are looking to add as a result of your edumarketing.

Your goal at the end of six months may be to increase your YouTube subscribers by 15%, add 200 new names to your CRM and increase your blog followers by 20%.

Whatever your goal is, be sure to know where you are starting and what you want to measure. If you are building a blog, then you want to have people subscribe to your blog or register for a downloadable you've created so you can build your CRM database.

Building followers will help them to see you as the expert and allow you to keep in touch with them.

Let's look at an example for our real estate agent:

Their first goal looks like this:

6 Month Goal
Identify 3 Topics
Identify 2 Content Headlines for each Topic
Develop 1 article, 1 video and 1 downloadable for each Content Headline
Have a total 18 Edumarketing pieces

Activity

Write Out Your 6 Month Goals

Next, let's see what those goals look like for our real estate agent. Their goal was to have three topics and two content headlines for each topic. Here's what that looks like:

Topic	Buying Investment Properties	Buying Your First Home	Buying a Condo
Content Headlines	5 Things to consider when buying an investment property	Saving for your First Home	How much are Condo HOA?
Content Headlines	How to find a good fix and flip	How much are you throwing away on rent?	When is the right time to downsize?

For each content headline, they will create three deliverables. By leveraging on each topic, they can easily create, produce and distribute content that is cohesive.

Your turn! Write out what your goals will look like if your goal is three topics and two headlines for each topic. Take some time to develop your 6-month goals.

Activity

Topic			
Content Headlines			
Content Headlines			

Next, determine your goals on what results you want to have from the content you've created.

For our real estate agent, their first edumarketing piece was to create a blog post, YouTube video and downloadable checklist for the "5 Things to Consider When Buying an Investment Property."

Their goal with this campaign would look like the following:

Topic	Content Headline	Delivery Channel	Deliverables	6 Month Goal
Investment Properties	5 Things to consider when buying an investment property	Blog Article	Begin by writing an article in Word on the 5 things to consider when buying a home. Will use links to sites like Forbes to back up information. Once article is written, post onto blog. Will be sure to add Google search words like investment properties, buying a rental home, etc.	Add 100 blog subscribers
		YouTube Video	Develop the video from the article. Will	Have 500 followers

103

| | | | record the video from the top 5 points in the article. The video will only be one minute long and can reference back to the article for more information. Video will be highlights of the article. | |
| | | Downloadable Checklist | Create a checklist for buying an investment property. Create a link to downloadable in CRM and capture email to build marketing database. | Add 200 names to database. |

You can create your goals based on each deliverable you develop or for all your content.

Let's look at what that looks like when our real estate agent sets their 6-month goals, taking into consideration all that they are doing for that 6-month period.

6 Month Goal
Add 200 followers to blog
Add 200 subscribers to YouTube Channel
Add 200 names to CRM database

Now, write your 6-month goal of what you want to add for your "fan" base.

Activity

Write out your 6-month goal of what you want to gain with your edumarketing.

Once you've completed your 6-month plans and goals, then you can move on and create it for 12 months, then 18 months. Yes, it takes some time to create your plan, but it is important to help you focus on what you are going to do and the results you can obtain by sticking with a plan. Think about it like planning out a vacation. You certainly would not head out on a vacation having no idea where you were going or having at least a general idea of what you will be doing. The average American spends 10–20 hours planning their vacation.[vii] Your business is surely worth at least that amount of time.

Now that we have a plan, let's look at delivering your content.

Chapter 14

MARKET, SHARE, PROMOTE AND BUILD

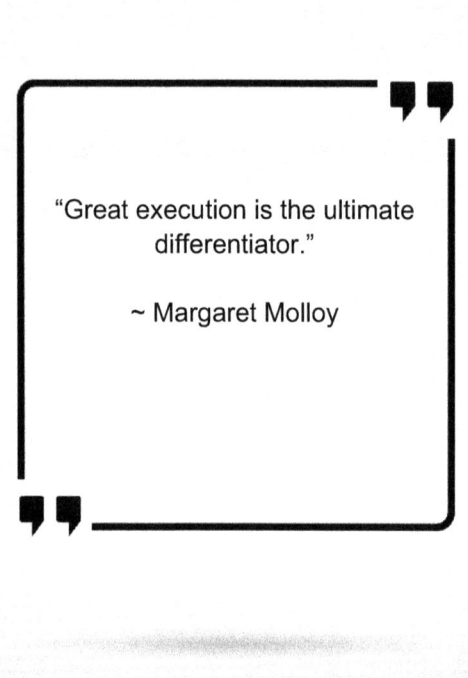

No matter how you've developed your content, you will be delivering your edumarketing content across a wide array of delivery channels. You may write a blog article on your website that goes out to your blog subscribers, but don't stop there. Take the link to that article and include it in your monthly newsletter to send out through your CRM or post it on LinkedIn or Facebook. You can send the link to your article to your business partners and ask if they would like to send it out to their database. If you have created a video for your YouTube channel, same thing. The more you leverage your content, the more you increase your visibility and your followers.

If you are doing a live workshop, be sure to promote a link for people to register for the workshop through social media. Take photos during the workshop and post those on social media. Use handouts during your live workshops that are downloadables that your participants can either give you their card and you can send them the items or have them go to your blog or website and register to receive the workshop handouts. Never give out your content without asking for at least their email. Your goal is to build your database, and one of the best ways to do that is by downloadables.

Create a plan that you always follow when you produce new content. Let's look at an example of what that looks like:

Delivery Channel	Facebook	LinkedIn	Blog	Partners	Blast Email
Blog Post	Post link to blog post	Post link to blog post	Write post	Email link to partners through CRM.	Include blog posts in monthly newsletter.
YouTube Video	Post link to YouTube Video.	Post link to YouTube Video.	Add YouTube Video link to blog post.		Include with monthly newsletter.
Downloadable Checklist	Only include if a standalone downloadable. Be sure to include call to action and ask to register.	Only include if a standalone downloadable. Be sure to include call to action and ask to register.	Should already be included in blog article. If not consider adding it as a blog post.	Only if applicable to partners.	Include in separate email depending on call to action on the downloadable.
Webinar	Promote on all social media. Target to audience if possible.	Promote on all social media. Target to audience if possible.	Include in online blog calendar on website.	Send to partners if related to their business.	Email to segmented list of prospects, partners or customers relevant to content.
Live Workshop	Promote on all social media. Target to audience if possible.	Promote on all social media. Target to audience if possible.	Include in online calendar on website.	Send to partners if related to their business.	Email to segmented list of prospects, partners or customers relevant to content.
Industry Publication	Post all publications.	Post all publications.	Post all publications.	Send any industry publication to partners.	Include in monthly newsletter or email as it relates to prospects or customers.
Tradeshow	Post to promote. Promote hard if speaking, presenting or participating in the event.	Post to promote. Promote hard if speaking, presenting or participating in the event.	Include in online calendar on website.	Send as it relates to partners.	Include monthly events calendar space in newsletter.

As you can see in this example, we have clearly identified steps to take to promote each of your edumarketing deliverables based on the delivery channel. The key is to know what you are doing so that you are consistent and can follow up for the best results. What good is it to have a blog post if you don't let anyone know about it? If you are doing events to showcase you as the "expert," you need to let people know that you are speaking at an event or on a webinar or have produced a new video. You can follow this example or create your own plan. Below is a blank template for you to use to complete your plan.

Delivery Channel	Facebook	LinkedIn	Blog	Partners	Blast Email
Blog Post					
YouTube Video					
Downloadable Checklist					
Webinar					

Live Workshop							
Industry Publication							
Tradeshow							

Now that you have your plan, you can create a checklist to make sure that you are doing what you should be doing to promote your edumarketing. If you have staff who are assisting you with your edumarketing, a checklist will help them to know when and what to do as well as create accountability. Once you have completed about 6 months of consistent edumarketing, you can start measuring your results, which we will talk about later.

Let's look at what that checklist looks like:

Date Published	Content Headline	Delivery Channel	Facebook	LinkedIn	Blog	Partners	Blast Email
9/15	5 Things to consider when buying an investment property	Blog	9/16	9/16	n/a	Emailed link in weekly email blast on 9/22	Emailed to investor list 9/16. Included link in monthly newsletter on 9/30
9/15	5 Things to consider when buying an investment property	YouTube	9/23	9/23	Posted link on 9/23	Emailed link in weekly email blast on 9/23	Emailed to investor list 9/16. Included link in monthly newsletter on 9/30

By using a planning checklist, you are able to manage your edumarketing plan as well as see what you are doing. There are also services available like Hootsuite that will allow you to plan your social media posts in advance, saving you time.

You can find the **Edumarketing Planning Workbook** at:

www.theedumarketer.com. Be sure to use the **code** edumarketer.

In the next chapter, I will share how to follow up with your prospects, measure your results and adjust your edumarketing plans.

Chapter 15

FOLLOW-UP AND MEASURING

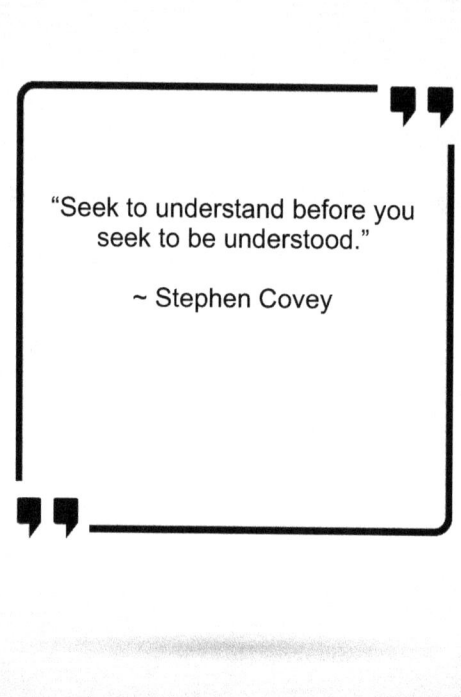

Sixty-three percent of people requesting information on your company today will not purchase for at least three months, and 20% will take more than twelve months to buy.[viii]

This is why creating an edumarketing plan is important. Edumarketing is not just creating something one time. Instead, it is creating an entire follow-up plan to educate, discover additional needs, measure the interest and effectiveness of your plan and adjust as needed.

Follow-Up

There are many ways you can follow up with your prospects. If you are doing a webinar, you can follow up by sending a copy of the information you covered. If you are posting a blog article, you can follow up by emailing out a checklist that relates to the information in the blog. If you are doing a live workshop, you can follow up with an additional course that goes into your topic in more detail. There are always ways to follow up with your edumarketing. One of the most effective means of follow-up is by email, so you should invest in a good email marketing system. There are several out there to choose from. Some of my favorites include Total Expert and Keap by InfusionSoft because they allow you to build journeys that will help build a solid edumarketing plan.

You should also follow up and measure your edumarketing campaigns to determine interest and effectiveness. If your blog article on "How to select a great baby-sitter" generated 20 new followers and was forwarded 5 times, you want to find out if the new followers are good potential prospects for you and your business. You should set up a follow-up campaign for new followers that welcomes them and introduces you to them. You don't want to scare them away, so you want to be subtle in your approach. Begin by thanking them for the follow. Let them know that you are committed to sharing helpful information. Give them something as a thank you for subscribing or following you. It could be an eBook or article. Maybe you could work with a local bakery and coffee shop and give them a coupon for a free cup of coffee on you. Engage with them and start the conversation. Don't immediately start selling to them. Remember, your edumarketing plan includes information on you and your expertise, so they will learn about you over time. After you have sent a few follow-up emails to your new prospects, you can move on to including one or two questions in one of your follow-up emails that allows you to find out a bit more about them. Remember not to do this at the very beginning of your follow-up. Your goal is to find out if your new prospects are good potential customers, so you will want to include some questions that relate to your service or product. If you are a real estate agent, you may ask if they are thinking about purchasing a home in the next 12 months or you may ask them if they would buy a new home or an older home. This will allow you to see where they are in terms of using your service.

If you've collected a phone number, maybe a quick phone call just to say hello and introduce yourself. Whatever it is that you do to follow up, be sure that you have a consistent follow-up plan and that your follow-up plan includes measuring the effectiveness of your edumarketing campaigns.

Let's look at what a good follow-up template looks like. We will use our example for our real estate agent.

Content Headline	Delivery Channel	Number of New Subscribers/Followers	Follow-up to New Subscribers/Followers	Total Number of Impressions/Views @ 30 days/60 days/90 days
5 Things to consider when buying an investment property	Blog	15	Dropped into new subscriber thanks for joining campaign, investment buying long term campaign and prospect long term campaign	257/568/896
5 Things to consider when buying an investment property	YouTube	78	Dropped into new subscriber thanks for joining campaign, investment buying long term campaign and prospect long term campaign	986/2,307/7,896

When you start measuring your content, you will quickly see what is working and not working and where to provide additional content based upon interest. You should also consider what you have done in terms of marketing to increase visibility of your content. For example, if you paid for Facebook ads for your YouTube video, you will want to note that in measuring your results. If you don't want to measure each content headline, you may want to use a monthly spreadsheet to track and measure your edumarketing.

Something like the example below will allow you to track your delivery channels and measure your results on a broader base.

	Jan	Feb	Mar	Apr	May	Jun	MoM Growth	Description
Blog	200	300	400	500	550	600	67%	Blog subscribers
Email	100	250	500	700	1,200	1,500	150%	Email addresses in the database
Facebook	100	500	800	1,200	2,500	3,700	370%	Facebook page likes
YouTube	100	300	750	975	1,500	1,875	187%	Corporate YouTube account followers
Webinars	36	78	197	310	420	520	%	Monthly educational webinar attendees.

There are several ways to track and measure your edumarketing. Find one that works for you and stick with it. The most important thing to remember is follow up with your new subscribers, measure your results and adjust your campaigns and channels as needed. You may find that your blog is not as effective as you thought it would be or you may be getting your best results with your YouTube videos; whatever works, you will only find out by measuring your results.

You can download your **Edumarketing Planning Guide** at:

www.theedumarketer.com

Be sure to use the book code **edumarketer** to receive the downloadable for this book.

Chapter 16

ADJUST YOUR PLAN, CREATE NEW CONTENT AND REPURPOSE

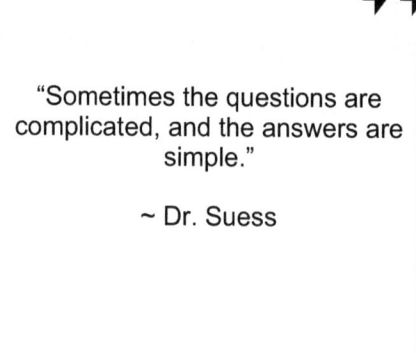

"Sometimes the questions are complicated, and the answers are simple."

~ Dr. Suess

Tracking your results is important so you know what is and isn't working. At the end of six months, evaluate your results and decide what you want to continue doing that is giving you the most results. Remember, it takes time to build an audience in some delivery channels. It also will take more than just one or two webinars to start building a following, so don't be too quick to stop trying something. If you are seeing great results in one particular channel or on one topic, then you want to pay attention and consider increasing your focus on that channel or develop additional edumarketing deliverables on that particular topic.

Your next step is to plan out your next 6 months. Follow our edumarketing planning guide to continue creating content. Remember to add additional channels now that you have developed some content. For example, you could take what you have developed over the past six months in a blog and turn them into articles or an e-book. You can take your videos and create an online course for your university. Don't be afraid to reuse content, especially if you have received positive results on it.

Let's review your planning and goal-setting steps:

1. **Develop Content**

Your first goal is to develop content. For example, your first goal may be to identify six topics, one for each month. From those six topics, you will write an article, produce a video and create a downloadable.

2. **Add Followers/Fan/Subscribers**

Your second goal is to increase your "fans," or followers. You know, the people who look to you for your expertise. The more followers you have, the more likely they will look to you at some time because you have established yourself as the expert. Sometimes prospects will convert immediately, depending on where they are in the buying cycle. For others, it may take some time before they become your customer. Staying top of mind with your expertise will help them to look to you when they are ready to purchase your product or service.

Consequently, you will want to also measure the number of "fans" you are looking to add as a result of your edumarketing.

Your goal at the end of six months may be to increase your YouTube subscribers by 15%, add 200 new names to your CRM and increase your blog followers by 20%.

Whatever your goal is, be sure to know where you are starting and what you want to measure. If you are building a blog, then you want to have people subscribe to your blog or register for a downloadable you've created so you can build your CRM database.

Building followers will help them to see you as the expert and allow you to keep in touch with them.

Activity

You already created your 6-month goals. Come back here at about 4 months and take time to develop your 12-month goals. Follow the same examples we used in Chapter 13 to create your 12-month goals. Use the charts we've provided for you in your **Edumarketing Handouts**. You can find them at **www.theedumarketer.com**

12 Month Goal

Determine your goals on what results you want to have from the content you've created.

You can create your goals based on each deliverable you develop or for all your content.

If you stay on top of your planning and goal setting and complete it every six months, you will be able to continue to develop new content, repurpose existing content and continue to build a following who sees you as the expert in your field.

We've covered a lot in the past chapters, so let's recap the steps to take to create your edumarketing plan:

Step One – Discover Your Expertise

Step Two – Identify What Your Customers Want to Know

Step Three – Create Your Edumarketing Plan

Step Four – Determine Your Delivery Channels

Step Five – Develop Your Content

Step Six – Create Your Message

Step Seven – Deliver Your Message

Step Eight – Follow Up

Step Nine – Measure Your Results

Step Ten – Adjust Your Plan/Create New Content

Step Eleven – Know When to Outsource

One of the biggest obstacles most professionals face is having the time and expertise to develop edumarketing content. You may be the top in your field as a Real Estate agent, but don't have the time or resources to develop content, post it on your blog and social media or create video content. This should not be what keeps you from developing an edumarketing plan. Unfortunately, however, it is what keeps most from utilizing this powerful marketing vehicle. If you are looking for help don't hesitate to reach to us at **www.edumarketing.com**.

Chapter 17

SETTING UP YOUR DELIVERY CHANNELS

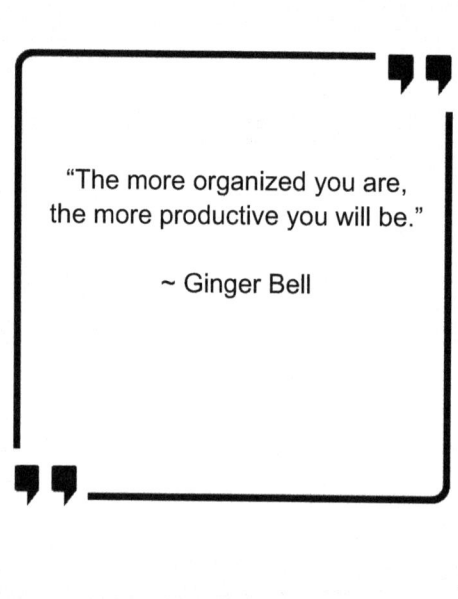

Choices, choices, choices. There are so many ways you can develop and deliver your edumarketing. We discussed these earlier, but I'll refresh your memory by listing them here again:

- Blogs
- Social Media: LinkedIn, Twitter, Instagram and Facebook
- Email Marketing Campaigns
- YouTube Channels
- TV Channels
- Podcasts
- Live Workshops
- Webinars
- Online University
- Email Courses
- Radio Shows
- Magazine and Newspaper Articles
- Whitepapers
- E-Books
- Print Books

The reason I recommend determining your delivery channels prior to developing your content is because you will develop different information based on your delivery channel. If you are a life coach who is going to create a YouTube channel to share "How to Build Confidence, Self-Worth & Self Esteem," you will want to map out your video series. You will list your topics for each video in the series. Then you will create an outline script detailing what you are going to talk about in each video. You will then record your videos and set up your YouTube channel and start marketing your series.

If you decide to start a blog for your website, you will want to create your topics, write your content, create your blog and map out a calendar to post your series.

If you are going to do a Podcast, you are going to create your topics, outline your talking points, maybe line up guest speakers and determine your delivery method.

If you are going to create an online course, whether it's an email course or one you place in your own online university, you will create your course topics and content, which may include lessons, handouts, resources, tools, etc.

Each delivery channel has different deliverables you will need to develop, which we will detail in the next chapter.

I recommend starting with the delivery channel that you feel will be the easiest for you to initially implement. If you are comfortable in front of the camera, then start with a video series. Maybe you decide to start with an email edumarketing campaign and find out what your customers want to know.

Whatever delivery method you decide on, you can always come back and re-create the same content in another delivery channel. You do not have to create content for each channel. In fact, you will be able to re-purpose your content to a variety of delivery channels.

Let's look at what you need to consider for each delivery channel.

This list of delivery channels by no means includes all that you can use for your edumarketing, but it gives you a good base.

BLOG

What You Need

Blogging Platform

If you already have a website, you can add a section for your blog. You can create sections on your website based on your customer segments. For a real estate agent, you may have a tab for First-time Home Buyers or Investors and when someone clicks on that tab, it takes them to a page that hosts your blog articles related to information for that customer. The great thing about blogs is that they can help to increase your ranking on search engines, so they are not only an effective edumarketing effort, but they can help to build your business, which is why we are doing this, right?

Lead Capture System

The most important items to include in your blog posts are the following:

1. **Call to Action** – this is why they should take the next step and contact you.
2. **Lead Capture Mechanism** – Offer them something that they want to learn about in exchange for their name and email. Trust me, people complete forms.

Email Follow-Up System

You've gotten them to share their information with you, now you need to follow up with them. You should build an email edumarketing campaign in a CRM platform like Keap, by Infusionsoft, Total Expert or Constant Contact that will drip additional information to them.

Steps to Take

Step 1: Create your plan by selecting your topic, creating an outline, conducting research, and checking facts. Remember, this is a series, not just one blog article. Your goal should be to create a long-term edumarketing plan, so you should have at least one general topic and then break that topic down into smaller segments. Each segment becomes a blog article.

Step 2: Craft a headline that is informative and will capture the readers' attention.

Step 3: Write your posts by either writing a draft in a single session or gradually working on parts of it. Again, it's a series.

Step 4: Use images to enhance your post. A picture speaks a thousand words and captures the reader's attention.

Step 5: Edit your blog post. Make sure to avoid repetition, read your post aloud to check its flow, have someone else read it and provide feedback, keep sentences and paragraphs short, don't be a perfectionist, don't be afraid to cut out text or adapt your writing last minute.

Step 6: Schedule your posts. If you have a WordPress site, you can write these all out at one time and then schedule them to post at a set time. Do this so that you only have to make blog posts once or twice a month.

Step 7: Remember to segment your topics based on your customer segments. People search for information online, and if they can easily find the information they are looking for on your site, they are more likely to continue reading.

Step 8: Include a call to action! This is the one area that most people miss out on. If you have someone on your site reading your information, give them a chance to subscribe for additional content. Don't put everything you want to teach them on your site. Instead, combine your blog with a whitepaper, checklist or top ten list. People LOVE lists, so add one to your blog and ask prospects to share their name and email address so you capture their data and continue the conversation.

Step 9: Email Follow-Up. If they have taken the time to give you their contact information for your free download, then they are an interested prospect and one that you should test for interest. Set up an email follow-up campaign in a CRM that will include short messages delivered over a period of time that will give them information and test for their interests.

MAGAZINE AND NEWSPAPER ARTICLES

Writing magazine and newspaper articles is a fantastic way to position yourself as the expert in your industry. Many trade publications are eager to receive good content from professionals in the industries they serve. You probably already subscribe to some magazines that serve your industry. You can also look to local publications who have readers who would be looking for the services you provide. Consider contacting your local chamber of commerce and finding out if they have a monthly newsletter you can write for. This is especially helpful if your business is that of a coach, trainer or speaker. If you are a real estate agent, you could write an article on the housing market in your area. There are tons of places to write articles. Make sure you get an online link to your published article and include it in your social media posts and email marketing campaigns!

Steps to Take

Step 1: Research organizations you can write for. Find out what their readers want to know about. Maybe you are an accountant and can write about recent tax law changes. Or maybe you are an elder attorney who can write about creating a will or trust.

Step 2: Write your article. If you are not keen on writing, then consider hiring a "ghost writer." You can hire a ghost writer who will interview you and then write your article based on your interview. There are several places to find good content writers on freelance sites like Upwork and FIVERR. Be sure to ask for examples of their work and references when you are using freelancers.

Step 3: Craft a headline that is both informative and will capture reader's attention.

Step 4: Write your bio. Most publications will not let you "sell" in your article, and that is fine, because the purpose of writing an article is to establish your expertise. Where you can "sell" yourself is in your bio. They may limit the number of words you can have, so be sure to let your readers know who you are, what you do, why they should call you and how to reach you.

WHITEPAPERS/E-BOOKS/CHECKLIST/TOP TIPS GUIDES

A whitepaper, e-book, checklist or top tip guide is a marketing tool. It is used to reach customers, as well as educate and build trust in you and your company. You can use a white paper to discuss an issue that your target audience wants to hear more about that shows you're a credible source on that topic. When done correctly, whitepapers, e-books, checklists and top tip guides provide education for your leads, customers and partners and, if used correctly, can lead them to do business with you.

Steps to Take

Step 1: Create your plan by selecting your topic, creating an outline, and creating your content. Remember, these are items you will include as giveaways or downloadables, so they may be used in various areas of your edumarketing campaigns.

Step 2: Craft a headline that is informative and will capture readers' attention.

Step 3: Write your content, either writing a draft in a single session or gradually working on parts of it. You can create multiple pieces for each topic. You may have a checklist of things they should do or a top ten list. Don't write super long articles. Make them short and impactful!

Step 4: Use images to enhance your articles.

Step 5: Edit your content. Make sure to avoid repetition. Read your content aloud to check its flow, have someone else read it and provide feedback, keep sentences and paragraphs short, don't be a perfectionist, don't be afraid to cut out text or adapt your writing last minute.

Step 6: Include a call to action! Don't forget to let them know why they should get in touch with you, and most importantly, how to reach you!

Step 7: Email Follow-Up. If they have taken the time to give you their contact information for your free download, then they are an interested prospect and one that you should test for interest. Set up an email follow-up campaign in a CRM that will include short messages, delivered over a period that will give them information and test for interest.

SOCIAL MEDIA: LINKEDIN, TWITTER AND FACEBOOK

What You Need

Social Media Platform

If you don't already have a business page set up on these social platforms, you need to. If you already have yours set up, be sure to check to make sure that you have good graphics and your contact information is correct. Also, it is good to include a brief description about your business that includes a call to action or reason they would look to you.

The great thing about using social media is that you don't necessarily have to create content specifically for social media. Instead, use it as a marketing vehicle to post all content that you are creating.

Simply post your blog posts or YouTube video links. This is a no-brainer.

EMAIL MARKETING CAMPAIGNS

Email marketing campaigns can be the base of everything you do in your edumarketing campaigns and one that you want to make sure you set up correctly. It all begins with using the right CRM platform. Seriously, don't skimp on this one.

What You Need

CRM Platform

If you don't have a CRM system, get one. CRM stands for Customer Relationship Management. The title says it all. It is a platform to manage your customers. It is not a platform to send out one dimensional email messages. If used correctly, a good CRM platform will be the hub of all your edumarketing efforts. It truly is one of the most important tools you will use in your edumarketing campaigns.

You may be in an industry that has industry-specific CRM platforms. These may be good to use because they have pre-built templates. If you are a life coach, business coach, small business owner, trainer, speaker or real estate agent, you cannot go wrong with Keap by Infusionsoft, Total Expert or Constant Contact.

Here is what you should look for in a CRM for your edumarketing campaigns:

Email Marketing Campaigns (Not Autoresponders)

Those old school email marketing tools are mostly based around simple auto responders that send out scheduled emails at planned intervals. But those emails get delivered to EVERYBODY on your list. This is not targeted marketing. It's old school, and it's quickly fading away. To create your edumarketing campaigns, you need to be able to create edumarketing email campaigns that target your prospects and help you deliver content targeted to the information they are looking for.

Marketing that Targets Behaviors

You want to use a CRM that will help you target behaviors. This means if you create an edumarketing message focused on instructions on a do-it-yourself candle and in your message, you have a link to order wax, then you want to know that customer is interested in buying from you. Same thing if you are a real estate agent who created an edumarketing campaign titled "Planning to Purchase a Home" and you have a handout in that message titled "Saving for your First Home." In this example, you would want to tag that customer as a first-time home buyer and send them messages that speak to being a first-time home buyer. Same thing for email opens, email clicks, and interest levels. The future is in behavioral-based edumarketing. The more targeted you get in providing relevant information to your leads, the more likely they are to see you as the expert and purchase from you.

Lead Capture System

The most important items to include in your CRM is a way to create lead capture landing pages that feed into your email marketing platform.

Easy Integration with Your Website

Many small businesses use WordPress for their website or blog. You want a CRM that will easily integrate or "plug-in" to your website. Infusionsoft plugs right into WordPress and even has some nifty enhancement features you can potentially program and quickly utilize.

Easy Way to Capture Data About Each Prospect and Customer

This is key. Your leads, prospects and customers are your greatest asset. You need a tool that can quickly and easily capture data about every single lead that comes in as well as existing customers, so you can then use that data intelligently to grow your edumarketing campaigns and your business.

All-In-One Solution

Unless you can afford an entire tech team of programmers and developers, you really need a system that can do it all for you. From processing credit cards to generating CRM sales reports, this is your ace in the hole. The more one system can do for you, the easier managing your edumarketing campaigns will be.

Let's move on and discuss the steps to take to create an email edumarketing plan.

Steps to Take

Step 1: Create your plan by selecting your topic, creating an outline, conducting research, and checking facts. Remember, this is a series of messages. Your goal is to create a long-term edumarketing plan, so you should have at least one general topic and then break that topic down into smaller segments. Each segment becomes an email campaign.

Step 2: Craft a headline that is both informative and will capture readers' attention.

Step 3: Write your emails, either writing a draft in a single session or gradually working on parts of it. Again, it's a series.

Step 4: Use images to enhance your emails. A picture speaks a thousand words and captures the reader's attention.

Step 5: Edit your emails. Make sure to avoid repetition. Read your post aloud to check its flow, have someone else read it and provide feedback, keep sentences and paragraphs short, don't be a perfectionist, don't be afraid to cut out text or adapt your writing last minute.

Step 6: Schedule your emails. If you are using a CRM like Keap by Infusionsoft or Total Expert, this is easy because you can set your edumarketing email campaigns to go out over a period of time. You can also create additional edumarketing paths for your prospects based on their interest. Once this is set up, the system takes over and not only communicates information directed specifically to your targeted prospects and customers but will let you know when they are looking for specific information so you can reach out to them directly.

Step 7: Include a call to action! Remember, your goal is to educate and inform and then get them to act. They cannot do this if you do not give them a call to action.

EMAIL COURSES

An email course is just an email autoresponder, which is a series of emails that are sent out over a period of time. While that may be true, an email course is more than that. If used in your edumarketing plan, an email course also teaches your audience something related to your business and their questions by delivering lessons over a series of automatically generated emails.

Each email becomes a mini lesson and at the end of the course, you have a call to action to ask for their business.

Email courses are a great edumarketing tool because an email course allows you to:

1. Collect email addresses
2. Educate your leads, prospects, customers and partners
3. Build trust
4. Market yourself as the expert

If done correctly, an email course is one of the easiest things to develop and include in your edumarketing plan.

What You Need

CRM Platform

If you have a robust CRM platform, you can create some cool email courses. The key to creating email courses that convert prospects to customers is to have short, interactive courses. This means you should have downloadable worksheets that the student can print off and complete, short videos they can watch, and quick and easy assignments. The goal is to drip information to them over a period of time rather than delivering everything all at once.

Steps to Take

Step 1: Create your plan by selecting your topic and creating an outline. Remember, this is a series of messages or mini lessons. Each lesson becomes an email campaign.

Step 2: Craft a headline that is informative and will capture readers' attention.

Step 3: Write your emails, either writing a draft in a single session or gradually work on parts of it. Again, it's a series.

Step 4: Use images to enhance your emails. A picture speaks a thousand words and captures the reader's attention.

Step 5: Edit your emails. Make sure to avoid repetition. Read your post aloud to check its flow, have someone else read it and provide feedback, keep sentences and paragraphs short, don't be a perfectionist, don't be afraid to cut out text or adapt your writing last minute.

Step 6: Schedule your emails. If you are using a CRM like Keap, by Infusionsoft or Total Expert, this is easy because you can set your edumarketing email campaigns to go out over a period of time. You can also create additional edumarketing paths for your prospects based on their interest. Once this is set up, the system takes over and not only communicates information directed specifically to your segmented customer but will let you know when they are looking for specific information, so you can reach out to them directly.

Step 7: Create a landing page to allow students to register or "opt-in" to your email course. This landing page can be used on your website, social media forums, YouTube forums, etc. Once you create a landing page, you can create a "vanity" URL, so you can direct them straight to that page. For example, if you are doing a live workshop for First-time Homebuyers, you may create a landing page that allows leads to register to take your "Buying Your First Home" course.

You could purchase a vanity URL for example: www.austinfirsttimehomebuyer.com. This makes it easy for your leads to find you. You could even create a postcard to hand out at the event or send an email to your attendees after the event that takes them to your landing page. The easier it is for someone to register for your course, the better results you will have.

Step 8: Include a call to action! Remember, your goal is to educate and inform and then get them to act. They cannot do this if you do not give them a call to action.

YOUTUBE CHANNELS

If you don't already have a YouTube channel for your business, it's a great time for you to set one up so you can realize the power of video content.

Video content marketing is one of the most cost-effective means to edumarket your business! Whether it's in the form of bite-sized how-to's or long-form tutorials, YouTube content represents a cost-effective way to educate your audience.

What You Need

Video Camera

You can go as professional or simple as you want in this area. Many of today's smart phones have incredible cameras, but remember, you want something professional. Consider investing in a camera to use to shoot video from your computer. If you don't want to be on camera, you can create informational videos using video editing software and a PowerPoint slide deck. There are a ton of options for creating video content on the YouTube channel.

Quality Microphone

Seriously, get a good one. There is nothing worse than not being able to hear the audio on a video. Blue Microphone offers some great affordable options. Make sure to test your audio quality before you record your entire video and get it right.

Video Editing Software

Camtasia is great for the novice video editor. It is affordable and easy to use. If you are an Apple fan, Final Cut is a fantastic software and if you are good at editing, you can use the Adobe software.

YouTube Channel

Steps to take to set up your YouTube channel:

Steps to Take

Step 1: Set up your YouTube Channel.

Before you can get started on YouTube, you're going to need a Google account.

You can either create a new, dedicated account specifically for your YouTube business channel or use an existing, personal account.

Creating a new login is often ideal as you don't have to worry about security issues tied to your personal Gmail.

As a side note, you don't have to use your business name when creating a brand channel account. YouTube gives you the option to add your business name after you have created your account.

Once your channel is set up, you can brand it and create playlists specific to your segmented categories.

Step 2: Create your plan by selecting your topic and creating an outline. Remember, this is a series of videos. You can create multiple playlists on your channel to market out to your segmented leads, customers and partners. You can also include recorded webinars into your YouTube channel.

Step 3: Craft a headline that is informative and will capture viewers' attention.

Step 4: Record your videos.

Step 5: Brand your videos and make them interesting to watch. Use images to enhance your videos if you are not doing a talking head video and include callouts in your videos if you want to convey a segment of your video.

Step 6: Edit your videos. Watch for lags in your speech or pauses. Add music to your videos so they are interesting.

Step 7: Upload your video(s) onto your YouTube Channel. Be sure to title and tag your video(s) and place them into playlists.

Step 8: Market your new videos. Placing your YouTube video links on social media, in your blog posts or in your email courses and campaigns are a great way to integrate video into all your edumarketing campaigns.

Step 9: Include a call to action! Be sure to post how the viewer can get in touch with you and your call to action in all your videos.

LIVE WORKSHOPS/SPEAKING SESSION

Yes, I know. For many of you, the last thing you want to do is to stand at the front of the room and give a presentation. However, using live workshops and speaking sessions is one of the most powerful tools you can use to edumarket your business! Here is why you need to include live workshops and speaking sessions in your edumarketing plan.

Live Workshops and Speaking Sessions Will Boost Your Credibility

Being the person sharing information, be it helpful tips, insights or industry updates, will build loyalty, trust and awareness for you and your company. Whether the audience knows you or knows nothing beyond your logo or your business name, hearing you speak will allow them to build a positive reference to you and see you as the expert. Connecting with your audience will establish rapport. Being the expert who is speaking at the front of the room allows you to make a connection with someone that you cannot achieve standing in your booth at a tradeshow or introducing yourself to a group at the local chamber. Once you have been at the front of the room, even if you are simply a panelist sitting on a panel of several speakers, you are the expert, and they will come and ask you questions. This is your place to demonstrate your knowledge. You can speak about what your own experience has taught you and share your advice. When you are on stage, people will look to you as the expert and you can build credibility and authority. Remember, you were invited to speak at this event because you are the expert on a respective topic or in a respective field. All you need to do is share your knowledge with your audience. It will build your reputation and, consequently, the reputation of your business. People tend to trust you more easily when they consider you as the expert, and that is what your edumarketing plan is all about!

Speaking Generates Leads

There are a few ways to show up at events:

1. **Attend** - You can show up as an attendee and watch what is going on at the event.
2. **Sponsor** - You can show up as a sponsor and show that you support what's going on at the event.
3. **Speak** - You can show up as a speaker and BE what's going on at the event.

People will probably walk up to you after you speak and ask you questions. Many people will recognize you after the session at the cocktail reception and they will come up and talk to you. Be sure to have cards with you and ask for their cards so that you can use the experience to build your database. You should also ask for a list of attendees so that you can follow up with those who attended your speaking event. You may want to do a handout and ask attendees to give you their cards so that you can follow up with them and send them the handout. Whatever you do, be sure to leverage your expertise and use speaking at conferences and events to grow your network. Sales often start increasing because of these efforts.

Workshops and Speaking Sessions to Educate Your Customers

By delivering presentations and engaging in public speaking, you can inform and educate your current and potential customers. If you've been invited to talk about your business, remember that the people sitting in the room are there because they do want to know more about a topic. This is your chance to answer questions, explain concepts, give them a better idea of what you are all about and basically draw a picture of your business the way you want it to be seen.

There are many benefits of public speaking for your business, from creating more and more personal publicity for your brand to expanding your network and making more sales. For this reason, public speaking should become one of your edumarketing tools.

Steps to Take

Step 1: Determine your topics. Before you start making a ton of phone calls to get your speaking calendar booked, begin with a list of topics you can speak on that you feel very comfortable with. The more comfortable you are on the topic, the easier it is for you to have a conversation.

Step 2: Create your plan by creating an outline and marketing information. Have a list of topics you can speak on; that way, when an association or someone calls to book you to speak, you can send them a list of topics they can select from. This makes it easier for them and more likely to ask you to come back and speak again.

Step 3: Craft a headline that is informative and will capture readers' attention.

Step 4: Make a list of possible speaking venues and start calling. Follow up with your list of topics and then call to keep in touch. It can sometimes take a while to be asked to be a speaker at some venues but keep at it because it is worth it.

Step 5: Create your content. You don't have to use a PowerPoint slide deck for all your presentations. In fact, some of the best presentations are simply Q&A sessions, town hall discussions or open forums. If you are going to go without a slide deck, be sure to outline your presentation so you stay on track and focused.

Step 6: Practice, practice, practice. Try to memorize the first few minutes of your presentations. Make it a solid opening and strong closing. If you memorize your first few minutes, you will come out strong and not have the um's and thought paralysis that often happens if you are not used to speaking in front of an audience.

Step 7: Be prepared. Be sure to have cards with you. Plan your 30-second elevator pitch that you are going to close with. Show up early and plan on staying late. Most importantly, dress the part. Even if the group is casual, you should still dress professionally. This doesn't mean you have to wear your best suit, but you need to look professional and be the best-dressed person in the room. Stand out in a good way.

WEBINARS

What You Need

Webinar Platform – GoToWebinar, Zoom or WebEx

Webinars have been around for over a decade. Online webinars and meetings have become the norm, and it is still one of the most effective edumarketing tools. You can use any type of "sharing" platform.

Product infomercials, How to's, Trends and Updates are great on a webinar. Consider doing a webinar series where you create a series of monthly webinars focusing on a trend title like Technology Trends or Compliance Trends. Then each month, take a new topic and discuss that topic. By doing this, you can keep the webinars shorter. For example, 15 minutes of content with 5 minutes Q&A and 3 minutes for the commercial. You may also consider doing panels on webinars because it creates synergy and allows you to build your database. Here's an example of how to do an expert panel webinar series:

Webinar Series

Here are some ideas:

Compliance Trends – Technology Trends – Housing Trends – Social Media Trends – Remodeling Trends (you get the picture, right?)

The first step in organizing your series is to put together a group of complimenting companies and experts. For example, if you are a mortgage professional, you can include an appraiser, escrow/title professional, remodeler, etc. Create panelists who would benefit by being involved in your series.

Begin by creating a series of 5-6 webinars, each with a different topic related to the trend or topic you have identified.

Next, write the copy and invitation for each webinar.

Finally, send your copy and invitation to the panelists you want to participate in the series. Ask them to be on the panel, tell them the date(s), topic(s), etc. Let them know that all they must do is show up and share their expertise. Make the PowerPoint presentation(s) and create the marketing. This makes it simple for you to have others included on your panels because all they must do is send their photo, logo, bio and one slide for a "commercial" during the webinar. Conduct a 30-minute "pow-wow" call a week or two before the webinar and discuss content with the panelists for that webinar. Create your slide deck based on these discussions and send the copy to each panelist for review.

Make it easy for them to be involved in your series. Many professionals would love to do webinars, they just do not know how or have the resources to do them. When you make it easy for them to be involved, they will be able to participate.

Be sure to add each company's logo to all the marketing material and send them a link to send out to—you guessed it—their database and social forums. Smart, right? When you involve others in your edumarketing, you can also build your database from their database.

Automated Webinar Platform

Automated Webinars are very similar to a live webinar in terms of edumarketing, but it is recorded. If interaction is not that important or if you are going to be doing your webinars repeatedly on the same topic, then an automated webinar platform is good to use.

Check out some examples of automated webinars before you dive in and start creating your own. There are a few more steps to creating automated webinars, like online storage of the video file and testing of registration process. Once you get it all down, it is a great lead generator.

Steps to Take

Step 1: Determine your technology. If you are doing an automated webinar, do your research and be prepared. If you are doing a live webinar, practice with the technology and know how to use it. Sit in on some online training of the technology you are going to use. Always get onto your webinar 15 minutes prior to start time so you can test everything before going live. There are things that can go wrong with technology.

Step 2: Create your topic and plan by creating an outline and marketing information. If you are doing a webinar series, then you want to create an outline and marketing information for each session in the series and market series.

Step 3: Craft a headline that is informative and will capture readers' attention.

Step 4: Make a list of possible partners who can join you as a panelist on the webinar.

Step 5: Create your content. You don't have to use a PowerPoint slide deck for the entire webinar, but you will need to have a slide deck for the opening and marketing slides. If you are doing a panel, you can do a Q&A session and put one question on each slide in your slide deck and ask it to the panel. Be sure to share all questions with the panelists first so that they can plan out their responses.

Step 6: Treat it like a radio show. Begin with a welcome, test for sound, explain how to ask questions and take time to build rapport with webinar attendees. Be sure to introduce yourself and your panelists. Remember, your audience cannot see you unless you are using video in your webinar, so connecting with them is important.

Step 7: Follow up. Record your webinars and send out the recording and any material as a follow-up to all attendees or repurpose it to use in other edumarketing campaigns like YouTube or in your Online University. The rule of thumb for webinars is that half of those who register will show up.

ONLINE UNIVERSITY

Creating your own online university is one of the most powerful distribution channels you can use in your edumarketing plan and one that is most underused. This is because it takes a bit of expertise to create an online university. However, once you have developed your own online university, you will wonder why you waited so long.

What You Need

LMS (Learning Management System) Platform – Tovuti, Kajabi, Wordpress

There are literally hundreds of LMS platforms out there to choose from. Instead of going through all the features, pricing, etc., we'll make it easy for you because we've worked with most of them. Our favorite platforms for edumarketing are only three:

1. Tovuti
2. Kajabi
3. Wordpress – LearnDash Plugin

Each of these platforms has its own unique features, but what we like about these platforms is their ability to easily create content. Each platform will host videos, text, questions, etc. Depending on your needs, you can decide which is best for you. If you need help, please do not hesitate to contact us at www.edumarketingAgency.com.

Steps to Take

Step 1: Determine your platform. You will want to spend a bit of time selecting your technology platform because you will not want to have to move everything in 6 months. Things you will want to look at are ease of use, cost, scalability, etc. You will also want to look at development time for the platform. If you select LearnDash with Wordpress, you will need to hire a website developer and allow time for site design. If you select Tovuti or Kajabi, you will have a much quicker time to market. Once you have selected your platform, you can move on to the next step.

Step 2: Site design. Remember, this is your online university where your prospects and clients will come to learn. Take the time to do it right and create a logo that speaks to your brand. Use colors that are appealing and write copy that will explain who you are, what you are doing and why someone would want to sign up for a course or community.

Step 3: Create your course list. You will want to have more than one course in your online university. Courses do not have to be long, but they do need to have some relative content. Take your topic list and create a series. You may explain the homebuying process, if you are a real estate agent. If you are an interior decorator, you can explain how to stage a home or paint a room. Whatever your expertise, an online university provides a delivery channel that showcases your expertise.

Step 4: Create your content. Video, audio, animated videos, text, questions, activities. You name it, you can create it in an online university, so have fun!

Creating online universities are our specialty, so give us a call and we can help you get started with your own branded edumarketing magnet with an online university!

You can find more about the services we offer at **www.edumarketing.com**

We've only covered a handful of your delivery channel options. Trends and technologies are constantly changing, and each industry has its own set of technology tools that are relevant. It's important to select the delivery channels you are going to start with and then give yourself time to build on each of them. Remember to add additional delivery channels as you repurpose your content over time.

Chapter 18

CREATING A CONTENT HUB

"Everyone needs to have 30 core audiences to create content for."

~ Gary Vaynerchuk

A content hub is a place where you can centralize your content into one location.

A Content Hub should be organized to hold your content planning spreadsheet, your written content, videos, slide decks, written articles, curated content, social media content, user generated content, documents and even lead forms.

Content management can be like herding cats. To generate leads and build engagement, you need two things: quality content and somewhere for it to live.

Putting all of your content in chronological order won't cut it, because valuable content will get buried quickly, and trying to go back and repurpose content will become a struggle. A content hub solves these challenges by giving you a logical structure to your content.

What Is a Content Hub?

The phrase "content hub" might seem overwhelming and send a chill down your spine. But trust me, it will make your life easier. First, let's take a second to understand what a content hub is. A content hub is a collection of content related to a topic. That's it! Not so scary now, huh? A hub is focused on specific topics and content, like a blog, but presents information in a more user-friendly manner.

A content hub allows you to organize content so that you can find the content and repurpose it later.

Steps to Creating a Content Hub

Step One:

Choose where you are going to organize your content. You may choose to use a service like Dropbox or Google Docs. Whatever you choose, you should select something that you can access anywhere. Dropbox and Google Docs can also allow you to give access to content to other people. This is important as you grow your edumarketing plans because you may have other people helping you create, organize and distribute your content.

Step Two:

Name your folders. You should have one folder for your content. Title your folder **Edumarketing Content or Killer Content** or **(Your Name) Content**. The name of the folder should be something that you can quickly recognize and is separate from your other folders.

Step Three:

Create and name content type folders within your primary folder. I like to number my folders so that the ones I want to access the most are organized at the top. I recommend the following folders:

1. Content Spreadsheet
2. Written Articles
3. Videos
4. Podcasts/Audio
5. Presentations
6. Handouts
7. Flyers
8. Ads
9. Social Posts
10. Headers
11. Photos
12. Music
13. Intro/Outro Video
14. Branding

You can always add more folders, but generally I like to have all of my content "types" together. You can also organize your folders by customers, topics or products and services. You may want to organize with the following folders:

1. Content Spreadsheet
2. Products
3. Topics
4. Prospects
5. Customers
6. Partners
7. Headers
8. Photos
9. Music
10. Intro/Outro Video
11. Branding
12. Logos

Whatever you choose, be sure to have a plan so you can easily add, find and repurpose your content.

Step Four:

Create and name secondary folders within your content type folders. Organizing folders within your content type folders will allow you to easily add, find and repurpose your content. If you are organizing your content by type of content, as in my first example, then you will want to create secondary folders that may look like this if you are a mortgage professional:

1. FHA Programs
2. VA Programs
3. USDA Programs
4. First-time Home Buyer
5. Investment Properties

Whatever your organizational strategy is, you want to also add folders for your headers, photos, music, intro/outro video, branding and logos. By creating folders, you will be able to quickly add these items to your content as well as quickly locate and repurpose your content.

Chapter 19

MANAGING YOUR PLAN

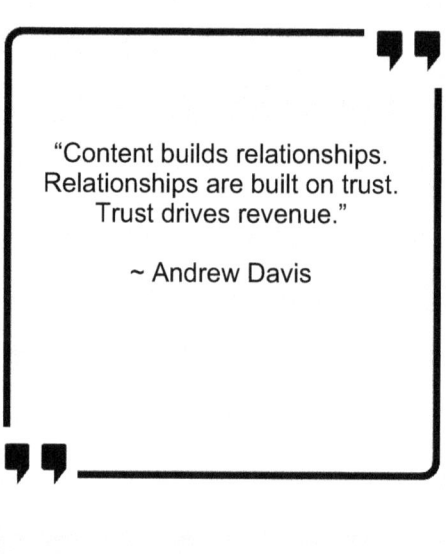

Creating a long-term edumarketing plan includes creating a system to manage your plan. It doesn't have to be complicated, but it does have to make sense for you. To get started on your management plan, let's recap what you have completed so far:

1. Identify your customers
2. Identify your customer segments
3. Identify your products and services
4. Identify your customers' problems
5. Identify your customers' questions
6. Identify your solutions to your customers' problems and questions
7. Create your topics
8. Write your headlines
9. Create your content
10. Identify your delivery channels

Managing this information may seem daunting, but it can be very simple. I develop and monitor edumarketing programs for companies with multiple audiences and multiple solutions, and I find the easiest way to manage their message, content and goals begins with a simple Excel spreadsheet. Sure, there are many project management programs available, but for most of you who are reading this book, you are an individual or small business owner who does not have a big budget or team. No worries. You got this. Let's begin by looking at what you should be including in your edumarketing management spreadsheet.

Up until now, we've used several planning tools, checklists and spreadsheets. Use these tools to flush out what you want to develop and then place your final plan into your management spreadsheet. Here are a few examples you can use to create your edumarketing long-term implementation and management plan. You will create a plan that fits your needs.

Example Number One

Topic	Customer(s)	Delivery Chanel(s)	Content Type	Delivery Date
Saving money for your first home	First-time home buyer	YouTube	Video	9/15

Example Number Two

Name	Purpose	Details	Audience	Form	Creation Date	Current Date	Location
Approved Investor List	General	List of approved Investors	Prospects, Clients, Investors	Publisher	2/6/2018	9/12/2018	Marketing and Sales-Shared Folder, Marketing, General Information FUEL Memos
Pricing Overview	General	Overview of current pricing	Prospects, Clients, Investors	Publisher	2/6/2018	2/6/2018	Marketing and Sales-Shared Folder, Marketing, General Information

Chapter 20

HOW TO GET MORE TRACTION WITH YOUR CONTENT

Think of your edumarketing content as your base. You create one piece of content that can then build into the next. You build your content from your expertise. It's the answers to your customers' questions, the solutions to their problems and the products and services you offer. The idea of a successful edumarketing plan is to take the content you create and leverage it among multiple delivery channels. If done correctly, you should be able to easily create and repurpose your content over and over again into videos, workshops, guides, posts, checklists, podcasts and more. Each step you take in repurposing your content builds your audience and following and supports the work you have placed into your edumarketing plan. You should not ever just publish one content item on your blog and be done with it. No, never!! Instead, you should have a plan in place to leverage your content across multiple mediums to gain the most possible reach.

Think of it like this:

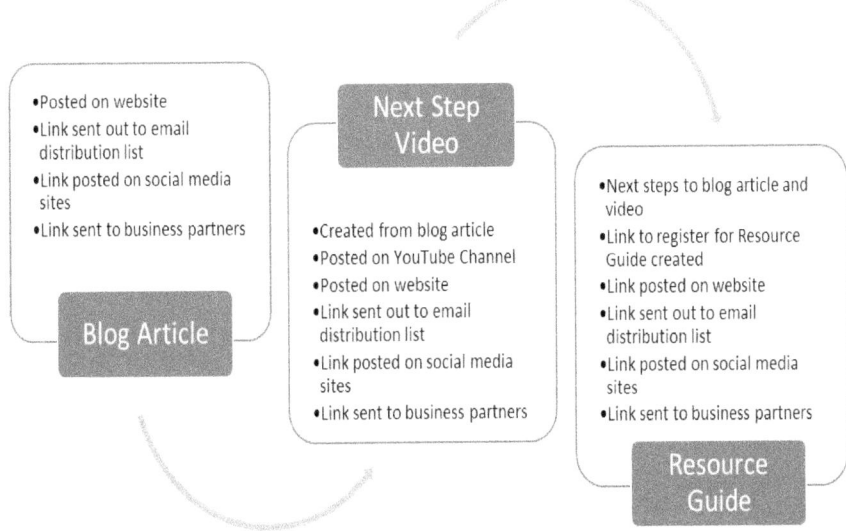

Most content is never used to its fullest potential. Many professionals simply go about creating a video post from time to time or writing an article to post in the local paper or putting together a workshop for their local chamber with no plan in place. Often it is done when business is at a low and they simply think, "Oh My! I'd better get my name out there." You are different, though, because first, you've invested in reading this information, and second, you now know the secret formula to creating an edumarketing plan. To get the most out of the content you create, you should follow these simple rules.

Create Evergreen Content

The questions you answer and problems you solve are not for a one-time audience, so creating evergreen content should be simple. If you are a realtor, your message may change based on some current trends, but the majority of what you create should be content that lasts for a long period of time and can be repurposed over and over again. Yes, the market may change, and you can add those particulars to some of your content, but the majority of your content should be able to live on for some time. Be sure to add information that speaks to current trends and industry insight but remember to keep much of your content evergreen so that you can repurpose it to speak to another customer segment or topic.

Know the Delivery Channels and How They Work Together

When you set out to write your edumarketing plans, you will focus on creating content for one or two distribution channels. You may start out creating blog posts that you can later repurpose into a video or expand upon for a checklist. If you are creating videos, you need to know how to post your videos on YouTube, Vimeo, Facebook or Instagram and how you can take those links and post them out onto other social media platforms, email and blog posts. Much of what you do in your edumarketing should be broadcast among multiple distribution channels. Take, for example, if you are speaking at a conference about "Five simple steps to long term financial freedom." You could create a video that highlights what you will be speaking on at the conference that you send to the conference planners to distribute to their distribution channel. You could then take that video and post it on your social forums, website, email distribution list, etc., and provide a link to register for the event. You could then take and create a handout that you will be giving to attendees at the conference after your session. The handout could be a checklist of what they need to do to take those "Five Steps." You could then post the handout after the conference to your social forums, website, email distribution list, etc., and share highlights of your session. You could also do a video that recaps the conference and send it to the conference planners and your distribution channels. You get the picture. Take each and every thing that you do and leverage it to get the most out of your edumarketing plan.

Understand Your Audience Segments and Speak to Their Specific Needs

Knowing your audience and their specific needs helps you to speak to them on a very personal level. Taking your content and adding in specific tidbits of information that speak to your segments will create a stronger bond with your target audience and establish the loyalty you are looking for as an expert. Take the time to add specific notes, information, and checklists that will help your segments find the answers and information they are looking for. Don't be afraid to be specific in your headlines and postings. No one ever appreciates spending their time reading through something that is only helpful to 20% of their needs. Be specific, and let your audience determine if it relates to them.

Have a System in Place to Organize and Manage Your Content

Not using a system to create, manage and repurpose your content is like throwing spaghetti against a wall and hoping that some sticks. I've never personally tried to do this but can imagine it is a difficult task. When you take the time to plan your edumarketing content, you will be able to produce and repurpose your content more quickly and efficiently. Take the time to do it right the first time. Even if you are not a planner, take the time to plan. It is well worth it. Remember to also have a good system to label and save your content so that you can easily go back and repurpose it.

Consistently Promote and Cross-Promote Content

You should never be afraid to reuse your content and cross-promote to other segments, products and delivery channels. Cross-promotion is a common form of marketing where customers of one particular product or service are targeted with promotion of a related product or service. Let's say, for example, that you are a dermatologist who has customers who have come in for precancerous skin conditions. You could write an article on the importance of protecting your skin from harmful sun rays and top sun damage therapies. One could assume that if they are concerned about their skin, they may want to have a treatment to repair damage. Knowing your customer segments, products and services will help you to create cross-promotion strategies.

Pay Attention to Your Audience and How They Respond to Your Content

Social media is all about likes, shares, and views, while your website is about tracking analytics, SEO, etc. How your audience responds to your content is important. It is not always about quantity, either. I've helped clients hold very small specific workshops, round table discussions and focus groups that have created lifetime customers and partners. It is important to have an idea of how your content is being perceived, and it can take some time to find out answers to that response. I recommend periodically asking questions to your social posts to check for relevance in your content. If you are holding webinars, conduct a survey at the end of the webinar and ask attendees what they liked, didn't like and would like to see.

I've had some of my best edumarketing programs come out of asking what people want in a next session after they just finished attending a session. If they are not responding, you may want to check to see that the content you are producing is relevant to their needs and questions. If not, change it or mix it up some. If you stick with what your customers are asking, you will usually be able to create content that your potential customers will respond to. The more you learn from your customers' reactions to your content, the better you'll be able to position yourself as the expert.

Create Modular/Rubber Band Content

As you are creating your edumarketing plan, you can think about how you will be able to repurpose your content. One of the easiest ways to create modular, or what I call rubber band, content is to create your content in modular formats. It's like taking each step and breaking it out into broader steps. This content in this chapter is a good example of modular/rubber band content. Each step I've listed out here is a module that can be expanded upon. Just like a rubber band stretches, so can your content. You may begin with a simple 5-part article that then stretches, and each of those 5 parts becomes another article, video or even lesson in your online university. I love using lists because it allows me to start with the basics and then build bigger pieces from those steps. Each step can be an article, video, webinar, workshop, or podcast. It really is limitless if you think about it, and you have so much information to offer. It is just a matter of beginning with the basics and then "expanding" from there.

Change Your Headlines, Images, Videos and Descriptions

Want more life from your content? Simply change the headlines, images, descriptions and videos. Freshen it up! Think about it like spring cleaning. If you are building a long-term edumarketing plan, you will eventually have a ton of content. Why not take advantage of that content and freshen it up with a new headline? No one will know that you've already created that content before if you create evergreen content. It's easy to change and share the information again. Be sure to keep the headline relevant to the content but add or decrease by a number. Take a list of 5 down to 3. Sometimes just adding a new photo and headline will get more views. It's also a great and easy way to build a larger content library.

Syndicate Your Content

There are other businesses related to yours who may serve the same type of customer without being a competitor. For loan originators, it's realtors, financial planners, insurance agents, attorneys, home remodelers, builders, etc. The content you create does not have to be yours exclusively. Consider becoming a guest blogger for a partner or for your local chamber. Offer your video content to a YouTube channel with more subscribers in exchange for some cross promotion. Syndicating your content is one of the easiest ways to build your following and gain additional exposure without having to do any additional work. Totally worth it!

Create a Monthly Newsletter

Yes, I know, newsletters have been around for a long time, but I still love them for one simple thing—consistency. Publishing a newsletter, whether online on your website, digitally by email or even by print, gets you consistently in front of the same people at least 12 times per month. You don't have to create brand new content, either. Simply include whatever you are working on for your monthly edumarketing. Also, consider asking others to contribute to your newsletter and in exchange send it out to their database. Newsletters also give you the opportunity to be seasonal. Consider adding tips that relate to what your readers are interested in. For example, include a list of firework shows prior to the 4th of July. Around Christmas, include a list of favorite Christmas light displays. You can include a top tips section, or did you know? The key to successful newsletters is a consistent template that readers can relate to. And…also be consistent. Send them out every month or quarter, and don't forget.

I encourage you to always be looking for ways to leverage your content and make it work for you.

Chapter 21

HOW TO SET UP AN EDUMARKETING STRATEGY ON A BUDGET

"We need to stop interrupting what people are interested in and be what people are interested in."

~ Craig Davis

There are a number of ways to maximize your edumarketing strategy for the best results, and you don't have to have a huge staff to develop, design and deliver your edumarketing content. Below are a few tips and strategies you can use to maximize your budget.

Start with a Plan

You can leverage your content by taking the time to actually create a plan. If you don't have a plan, you will end up moving from one distribution to channel to another without building a consistent message with valuable content. Take the time to write your plan. Get the edumarketing planning workbook and go through the process of identifying your expertise, defining your customers, discovering your customers' questions, problems and the solutions you offer. Build out on your topics and headlines and build your content plan in bigger modules that will break into smaller bite-size content. Then take your plan and identify your delivery channels and implement your plan.

Think Beyond Today

Build your edumarketing plan with tomorrow in mind. Think about where you want to take your business. Think about what your customers are asking today and how you can help them find solutions for tomorrow. Use a content grid to develop content that will build into tomorrow and answer long-term questions for your customers. Build yourself as the trend setter who knows what's happening and why your potential customers, customers and partners want to look to you as the expert in your field. It's all about building yourself up as the expert.

Batch Your Content

Batching your content goes hand-in-hand with your planning. Schedule a time each week to create your content. Block out time when you can be clear minded and committed to creating your content. Schedule out 5-7 days of content so you only have to do it one time and can easily push the content out on a schedule. Think of it like making a cake. You wouldn't go to the store to buy eggs and bring them home and then go back to the store and buy flour and then bring it home and then go back to the store and buy oil. No, instead you find a recipe, create a shopping list, go to the store, get all the ingredients, come home and follow the recipe to make the cake. Same holds true for your edumarketing content. Follow your plan and set aside time when you are going to create your content. Depending upon your delivery channel, there are many automated systems you can use to publish your content once and then push it out when you want. Be sure to also create a system to manage your content. I've given you a few examples. Organize your content like you would your cupboards. You don't have your dry foods mixed in with your silverware. Create a system and use the system.

Have a Response System

Plan on knowing how you are going to respond to likes, questions, and interests after you have posted your content. If you are doing a live workshop, you should have a way to follow up with attendees. Make sure you are engaging them after the conference. Get them to register for a free downloadable so that you can continue to keep in touch with them. Do a giveaway and collect cards and scan their cards into your CRM and tag them for follow-up. So many times, I see people give a presentation and walk away with absolutely nothing. I'm not recommending you spam them, but you are speaking for a reason. Leverage your time and don't be afraid to provide them with additional information.

Promote Your Edumarketing Content Through Email

Use the content you are creating in your newsletters or send out emails to your segments letting them know that you have created a checklist, video or helpful guide that they may be interested in. Don't be afraid to take your online information offline. I like to make phone calls to my prospects and let them know that I just wrote an article or did a video on a topic that they might be interested in. You can simply leave a message if you like and tell them that you are going to be emailing them a link to an article you just wrote that you thought they might be interested in. Often, a personal phone call will be just what they need to recognize your email and click on the link. Follow up with people who click on your videos, ask questions or post comments. Be sure to respond to every comment and do so in a timely manner. Thank them for their comments, too.

Promote Your Content on All Your Delivery Channels

No, you may not want to promote everything everywhere, but make sure that you are cross promoting your content. If you post a video on your YouTube channel, be sure to post that link out to your social media channels. You can use a platform like Hootsuite to manage your social media postings with one click. Be sure to include a Call to Action (CTA) at the end of each video, post or blog. It doesn't have to be salesy or long, but it does need to include additional information on how to reach you.

Use Visuals in Your Content

You will get more out of your content by adding photos and videos to your content. People scan for photos and headlines before they will read a lengthy article with a whole lot of words. Be sure to make your content visually appealing.

The key to anything is time, and although it may not take a lot of time to create an effective edumarketing plan, it does take planning and time. If done correctly, you can easily create a cost-effective edumarketing plan that can produce serious results for you and your business.

Chapter 22

KNOWING WHEN AND HOW TO OUTSOURCE

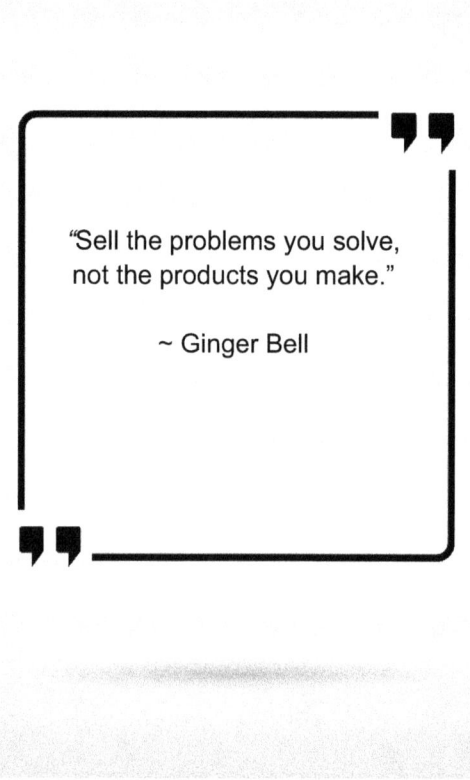

You're busy running your business, managing your staff, and making sales and the thought of adding something else is completely overwhelming. You know you need to do marketing and you believe that edumarketing is the answer. You want to position yourself as the expert in your field and you've got content to share, but you have absolutely no time to dedicate to building a plan, creating content and managing your long-term strategy and plan. I get it, and that's how I started my business. My clients did not have the time, resources, knowledge and expertise to create and manage their educational marketing. That's why they hired me to do the job for them. That's why I'm able to pass on this information to you, because I've learned from the trenches what works and what doesn't. I know the value of taking your expertise and building an edumarketing plan. If you are one of those people who doesn't want to do the heavy lifting yourself, don't feel alone. I think that people should do what they are best at and hire or contract out the rest. Even if you have an internal marketing department, they may not know what it takes to create an educational program. The first step in identifying if you need help and should outsource is to answer this simple question:

"Do you or someone on your team have the experience or willingness to learn what you need to do to properly plan, create, develop and deliver your edumarketing plan?"

If the answer is a resounding yes, then you can move on. If the answer is; not sure, then you may want to consider outsourcing some or all of your edumarketing.

There are several reasons a company may choose to outsource some of your edumarketing activities. They may include:

1. You are an individual entrepreneur, small team or company who does not have resources for marketing.
2. You are a team or company whose existing marketing team is lacking in some of the skills necessary to implement your edumarketing program.
3. You are a small team or company that needs some outside advice on what to focus on for your edumarketing plan.
4. You are a small team or company that has specific niche products that needs extra assistance in content creation.
5. You are an individual entrepreneur, small team or company who has tried to implement edumarketing unsuccessfully in the past.
6. You are a small team or company who recognizes your strengths and are committed to outsourcing to the strengths of others in their fields of expertise.

You can outsource some or all of your edumarketing activities. Below are some of the processes to consider outsourcing:

1. Strategy and Planning
2. Content Development
3. Content Delivery
4. Email Marketing
5. Blogging/Copywriting
6. Social Media Management
7. Video Production
8. Electronic Newsletters
9. eBooks
10. Guidebooks, Handouts
11. Presentation
12. Event Management
13. Photo, Audio Voiceover and Video Production
14. SEO
15. Course Development
16. E-learning Implementation and Management

Different projects require different skillsets. If you have never outsourced, you may be reluctant to do so because you think it costs too much or is too hard to manage. Maybe you don't want to let go of the management of the project. Based on my experience, you should look at outsourcing at least some of your edumarketing activities. Especially in the planning and e-learning implementation and management functions. What exactly are the pros and cons to outsourcing?

Pros to Outsourcing

Cut Costs

Outsourcing can actually save you money because when you contract with an individual or company, you do not have to pay the cost of taxes and benefits that you would if you hired a full-time employee. If you hire a content blogger, you only pay for the services they perform. Plus, you are able to expand as you need in your planning, so you can start out with only a few articles a month and then as you grow you can add additional articles.

Get Quality Content

When you contract for a skillset, you can expect quality results. Be sure to ask for examples of someone's work and don't be afraid to test a few people or companies out before you find what you are looking for. You may find a local videographer who is awesome and can meet your needs by coming to your office every Tuesday to do videotaping. You may find that resourcing through platforms like Fiverr works for you. Take your time and find the right fit for you and your needs.

Less Management

Most contractors are accustomed to managing large projects themselves, so you don't have to spend time managing them. You also don't have the headache of office drama and politics. The key is to find the right contractor to either work in the area of expertise you need, like writing or social media management or vlogging, or have one company manage all of your planning and execution.

Cons to Outsourcing

Risk of Hiring Wrong Contractor

Just like you can make a bad hire for an employee, you can make a bad hire of a contractor. You will want to manage your first part of your project and set expectations, timelines, budgets and communication methods. The more you set expectations at the beginning, the easier it will be. Don't be afraid to fire someone quickly if need be. Always be sure to set up a system for you to receive the content they are working on. Don't wait to the end for delivery. Manage it each week, and if they don't meet your expectations, let them know what you do expect.

Risk of Ruining Your Reputation

Yes, you still run this risk. Especially if you let someone manage your social media. A missed comment or inappropriate response that doesn't speak to you or your brand could cause you a ton of pain. Be sure to check references of those you are hiring. Really check them, too. Call who is on their list and make sure you look up the phone number and don't rely on the number they give you. Do your homework upfront. Especially if you are letting them loose on your brand.

May Not Be Easily Accessible

This goes back to setting expectations. Make sure you know the hours they are working. Outline timelines and set expectations. Some web developers, writers and bloggers work late hours and may not be accessible when you conduct your business. Find out what hours they will be available and set the standard. Don't be afraid to conduct weekly check-in calls, especially as you are getting started on a project.

Outsourcing with Interns

Some of my best contractors have been with college students who act as interns. You can decide whether or not you want to pay them. I always believe in paying someone something for their work. I find that it gives me better engaged and interested interns. You may find that you don't mind managing interns and can create a "learning" internship where you help teach them about your business in exchange for helping you create videos or manage your social media. Check with your local community colleges and schools to find out what programs they have and if there are students who would be interested in or may benefit from working with you in your business. Just as in outsourcing to professionals, you should set expectations with your interns. Let them know when they are expected to work, what they will be working on and what's in it for them. There are lots of students out there who are looking for a chance, and your opportunity may just be a perfect match for both of you.

Having worked with many companies both large and small, I recommend outsourcing at least some of your edumarketing activities. It may be something as simple as your video editing and production. Just be sure to make good use of your time and create and produce your content to your expertise.

Chapter 23

IT'S A MARATHON, NOT A SPRINT - CREATING A LONG-TERM STRATEGY

> "The aim of marketing is to know and understand the customer so well the products or service fits him and sells itself."
>
> ~ Peter Drucker

Congratulations! You have completed this book and are on your way to creating an edumarketing plan using your expertise. If designed properly, it will be the base of everything you do in your business and can create long-lasting results. The key is to be intentional and have a plan, so you don't suffer from analysis paralysis. It does take planning and time. You wouldn't head off for a two-week vacation without having an idea of where you are going, what you are going to do, where you are going to stay and how much fun you are going to have. The same holds true for your edumarketing plans. The best time to start is now! And remember, when it comes to creating your strategy, you need to create a plan that will give you and your team the ability to build on your expertise, remain organized and proactively develop content. Over time, you will be able to develop a robust library of content positioning you as the expert. What you do now will affect your business today and tomorrow!

To create a long-term strategy, be certain to take the following steps:

1. Identify your expertise
2. Understand your customer segments
3. Create a list of common customer questions, solutions and solved problems you provide for your customers
4. Make a list of topics that will help you create evergreen content
5. Write your headlines
6. Develop your content
7. Organize your content
8. Create a delivery schedule
9. Make a plan to create modular and rubber band content you can quickly pull from and add to
10. Most importantly, make your content personal and have fun!!

When creating your long-term strategy, be sure to include your yearly content initiatives that you can use in your newsletters and look for partners to help you create additional content.

Remember to include multiple delivery channels in your edumarketing plan like workshops, conferences, webinars, social media, handouts, check lists, etc.

Build your database using your CRM and utilize your email communication to include things like weekly/monthly advice, promotions of webinar series, industry events and workshops.

Our world of technology, social media and information is rapidly changing, and I encourage you to visit my site edumarketing.io for the latest in updates and tips.

Also, I'd love to hear your feedback, successes and comments.

Want more Edumarketing?

Visit us online at **www.edumarketing.com**

Ready to learn more about how to create your own edumarketing plans? Check out our Edumarketing Academy at **www.edumarketingacademy.com**

Would you like help creating your content, videos or edumarketing plan? We've created our Edumarketing Agency for clients who would rather stick to what they know and let us position them as the experts they are. Connect with us and see how we can help you at **www.edumarketingagency.com**

In the meantime, I wish you the very best in all that you create! Let's go edumarket!

To the edumarketer in **YOU**!

~ Ginger Bell

#theedumarketer

#edumarketing

#edumarket

Connect with Ginger Bell on Social Media

https://www.facebook.com/edumarketingagency

https://www.twitter.com/gingergbell

https://linkedin.com/in/gingerbell

Chapter 24

USEFUL TOOLS, TECHNOLOGY AND RESOURCES

"The only thing that 100% connects everyone is that we need someone's attention before we can tell them what we want."

~ Gary Vaynerchuk

Below are links too various resources that you may find helpful in your edumarketing planning, development and implementation. Links to these resources does not constitute an endorsement or recommendation. These resources are meant to only be a helpful guide.

The Edumarketer Book Download

Be sure to head over to our website to get your downloads for the book. We've taken everything we talked about in this book and created an Edumarketing Planning Workbook to help you get started. Just go to the website below click on book handouts. Enter your information and use the code "edumarketer" to get your book downloads.

www.theedumarketer.com

Code: edumarketer

Project Management Tools

Asana

Basecamp

Monday

Trello

Workzone

Outsourced Solutions

EdumarketingAgency (Yes, we are available for hire! Connect with us at **www.edumarketingagency.com** and let us know your project needs.)

Fiverr

Contently

WriterAccess

Scripted

Upwork

Graphic Design Resources

Adobe

Canva

PicMonkey

Design Pickle

Flocksy

CRM/Marketing Operating Systems

Total Expert

InfusionSoft by Keap

Constant Contact

Online Learning Platforms

Tovuti

Kajabi

LearnDash

Stock Photos Sites

Unsplash

StoryBlocks

iStock

ShutterStock

Pexels

StockPhoto

Stock Audio Sites

AudioBlocks

Premium Beats

EpidemicSound

AudioNetwork

iStock

Stock Video Sites

Video Blocks

iStock

ShutterStock

Videvo

Video Editing Software

Adobe Premiere Elements

Adobe Premiere Pro

Apple Final Cut Pro X

Apple iMovie

Camtasia

Corel Video Studio

CyberLink PowerDirector

Pinnacle Studio Ultimate

Magix Movie Edit Pro

Nero Video

Wondershare Flimora

Animated Video Maker Software

Animaker

Moovly

Powtoon

Video Email Software

BombBomb

Loom

Webinar/Meeting Software

Gotomeeting

Gotowebinar

Join.me

WebEx

Zoom

Automated Webinar Software

Easy Webinar

Stealth Seminar

WebinarJam

ABOUT THE AUTHOR

Ginger Bell is the founder of Edumarketing.com, Edumarketing Academy and Edumarketing Agency, a full-service education, marketing, video production and business management firm focused on using the power of education, new technologies, media and personality-driven marketing to position individuals and companies as experts in their field.

Ginger has co-authored two best-selling books with Brian Tracy titled "Cracking the Success Code" and "Success Today. She has also co-authored a best-selling book with Jack Canfield titled, "Success Breakthroughs", where she writes about mastering the art of edumarketing. Ginger has been awarded the Book Marketer of the Year award by the Association of Experts, Writers and Speakers as well as two Expy Awards for the education programs she has developed within the mortgage industry.

A former Dale Carnegie Training Consultant, Ginger has consulted with companies such as RE/MAX, Motto Mortgage, First American Title, Finance of America, FirstFunding, Arch MI and countless entrepreneurs and small business owners across the U.S.

A sought-after speaker and published author nationwide, Ginger has been named as one of Mortgage Banking's Most Powerful Women, 50 Most Connected Professionals and Elite Women of Lending. Ginger has been awarded the "Professional Woman of the Year" award by the National Association of Professional Women for her commitment to training and education. Ginger uses edumarketing with her clients to help them position themselves as experts in their field.

Ginger lives in Portland, Oregon with her husband, Blair and her cat, Gigi.

---

[i] https://en.wikipedia.org/wiki/Education
[ii] http://www.businessdictionary.com/definition/marketing.html
[iii] https://en.wikipedia.org/wiki/Expert
[iv] https://www.radicati.com/wp/wp-content/uploads/2015/02/Email-Statistics-Report-2015-2019-Executive-Summary.pdf

[v] https://www.statista.com/statistics/456500/daily-number-of-e-mails-worldwide/
[vi] https://www.statista.com/statistics/323878/us-monthly-minutes-online-video-age/
[vii] https://www.quora.com/How-much-time-does-the-average-American-spend-researching-and-planning-a-domestic-trip-vacation

[viii] https://www.marketingdonut.co.uk/sales/sales-techniques-and-negotiations/why-you-must-follow-up-leads

www.ingramcontent.com/pod-product-compliance
Lightning Source LLC
Chambersburg PA
CBHW060846170526
45158CB00001B/257